MONOGRAPHS ON STATISTICS AND APPLIED PROBABILITY

General Editors

D.R. Cox, V. Isham, N. Keiding, T. Louis, N. Reid, R. Tibshirani, and H. Tong

1 Stochastic Population Models in Ecology and Epidemiology *M.S. Barlett* (1960)
2 Queues *D.R. Cox and W.L. Smith* (1961)
3 Monte Carlo Methods *J.M. Hammersley and D.C. Handscomb* (1964)
4 The Statistical Analysis of Series of Events *D.R. Cox and P.A.W. Lewis* (1966)
5 Population Genetics *W.J. Ewens* (1969)
6 Probability, Statistics and Time *M.S. Barlett* (1975)
7 Statistical Inference *S.D. Silvey* (1975)
8 The Analysis of Contingency Tables *B.S. Everitt* (1977)
9 Multivariate Analysis in Behavioural Research *A.E. Maxwell* (1977)
10 Stochastic Abundance Models *S. Engen* (1978)
11 Some Basic Theory for Statistical Inference *E.J.G. Pitman* (1979)
12 Point Processes *D.R. Cox and V. Isham* (1980)
13 Identification of Outliers *D.M. Hawkins* (1980)
14 Optimal Design *S.D. Silvey* (1980)
15 Finite Mixture Distributions *B.S. Everitt and D.J. Hand* (1981)
16 Classification *A.D. Gordon* (1981)
17 Distribution-Free Statistical Methods, 2nd edition *J.S. Maritz* (1995)
18 Residuals and Influence in Regression *R.D. Cook and S. Weisberg* (1982)
19 Applications of Queueing Theory, 2nd edition *G.F. Newell* (1982)
20 Risk Theory, 3rd edition *R.E. Beard, T. Pentikäinen and E. Pesonen* (1984)
21 Analysis of Survival Data *D.R. Cox and D. Oakes* (1984)
22 An Introduction to Latent Variable Models *B.S. Everitt* (1984)
23 Bandit Problems *D.A. Berry and B. Fristedt* (1985)
24 Stochastic Modelling and Control *M.H.A. Davis and R. Vinter* (1985)
25 The Statistical Analysis of Composition Data *J. Aitchison* (1986)
26 Density Estimation for Statistics and Data Analysis *B.W. Silverman* (1986)
27 Regression Analysis with Applications *G.B. Wetherill* (1986)
28 Sequential Methods in Statistics, 3rd edition
G.B. Wetherill and K.D. Glazebrook (1986)
29 Tensor Methods in Statistics *P. McCullagh* (1987)
30 Transformation and Weighting in Regression
R.J. Carroll and D. Ruppert (1988)
31 Asymptotic Techniques for Use in Statistics
O.E. Bandorff-Nielsen and D.R. Cox (1989)
32 Analysis of Binary Data, 2nd edition *D.R. Cox and E.J. Snell* (1989)

33 Analysis of Infectious Disease Data *N.G. Becker* (1989)
34 Design and Analysis of Cross-Over Trials *B. Jones and M.G. Kenward* (1989)
35 Empirical Bayes Methods, 2nd edition *J.S. Maritz and T. Lwin* (1989)
36 Symmetric Multivariate and Related Distributions
K.T. Fang, S. Kotz and K.W. Ng (1990)
37 Generalized Linear Models, 2nd edition *P. McCullagh and J.A. Nelder* (1989)
38 Cyclic and Computer Generated Designs, 2nd edition
J.A. John and E.R. Williams (1995)
39 Analog Estimation Methods in Econometrics *C.F. Manski* (1988)
40 Subset Selection in Regression *A.J. Miller* (1990)
41 Analysis of Repeated Measures *M.J. Crowder and D.J. Hand* (1990)
42 Statistical Reasoning with Imprecise Probabilities *P. Walley* (1991)
43 Generalized Additive Models *T.J. Hastie and R.J. Tibshirani* (1990)
44 Inspection Errors for Attributes in Quality Control
N.L. Johnson, S. Kotz and X, Wu (1991)
45 The Analysis of Contingency Tables, 2nd edition *B.S. Everitt* (1992)
46 The Analysis of Quantal Response Data *B.J.T. Morgan* (1992)
47 Longitudinal Data with Serial Correlation—A state-space approach
R.H. Jones (1993)
48 Differential Geometry and Statistics *M.K. Murray and J.W. Rice* (1993)
49 Markov Models and Optimization *M.H.A. Davis* (1993)
50 Networks and Chaos—Statistical and probabilistic aspects
O.E. Barndorff-Nielsen, J.L. Jensen and W.S. Kendall (1993)
51 Number-Theoretic Methods in Statistics *K.-T. Fang and Y. Wang* (1994)
52 Inference and Asymptotics *O.E. Barndorff-Nielsen and D.R. Cox* (1994)
53 Practical Risk Theory for Actuaries
C.D. Daykin, T. Pentikäinen and M. Pesonen (1994)
54 Biplots *J.C. Gower and D.J. Hand* (1996)
55 Predictive Inference—An introduction *S. Geisser* (1993)
56 Model-Free Curve Estimation *M.E. Tarter and M.D. Lock* (1993)
57 An Introduction to the Bootstrap *B. Efron and R.J. Tibshirani* (1993)
58 Nonparametric Regression and Generalized Linear Models
P.J. Green and B.W. Silverman (1994)
59 Multidimensional Scaling *T.F. Cox and M.A.A. Cox* (1994)
60 Kernel Smoothing *M.P. Wand and M.C. Jones* (1995)
61 Statistics for Long Memory Processes *J. Beran* (1995)
62 Nonlinear Models for Repeated Measurement Data
M. Davidian and D.M. Giltinan (1995)
63 Measurement Error in Nonlinear Models
R.J. Carroll, D. Rupert and L.A. Stefanski (1995)
64 Analyzing and Modeling Rank Data *J.J. Marden* (1995)
65 Time Series Models—In econometrics, finance and other fields
D.R. Cox, D.V. Hinkley and O.E. Barndorff-Nielsen (1996)

66 Local Polynomial Modeling and its Applications *J. Fan and I. Gijbels* (1996)
67 Multivariate Dependencies—Models, analysis and interpretation
D.R. Cox and N. Wermuth (1996)
68 Statistical Inference—Based on the likelihood *A. Azzalini* (1996)
69 Bayes and Empirical Bayes Methods for Data Analysis
B.P. Carlin and T.A Louis (1996)
70 Hidden Markov and Other Models for Discrete-Valued Time Series
I.L. Macdonald and W. Zucchini (1997)
71 Statistical Evidence—A likelihood paradigm *R. Royall* (1997)
72 Analysis of Incomplete Multivariate Data *J.L. Schafer* (1997)
73 Multivariate Models and Dependence Concepts *H. Joe* (1997)
74 Theory of Sample Surveys *M.E. Thompson* (1997)
75 Retrial Queues *G. Falin and J.G.C. Templeton* (1997)
76 Theory of Dispersion Models *B. Jørgensen* (1997)
77 Mixed Poisson Processes *J. Grandell* (1997)
78 Variance Components Estimation—Mixed models, methodologies and applications
P.S.R.S. Rao (1997)
79 Bayesian Methods for Finite Population Sampling
G. Meeden and M. Ghosh (1997)
80 Stochastic Geometry—Likelihood and computation
O.E. Barndorff-Nielsen, W.S. Kendall and M.N.M. van Lieshout (1998)
81 Computer-Assisted Analysis of Mixtures and Applications—
Meta-analysis, disease mapping and others *D. Böhning* (1999)
82 Classification, 2nd edition *A.D. Gordon* (1999)
83 Semimartingales and their Statistical Inference *B.L.S. Prakasa Rao* (1999)
84 Statistical Aspects of BSE and vCJD—Models for Epidemics
C.A. Donnelly and N.M. Ferguson (1999)
85 Set-Indexed Martingales *G. Ivanoff and E. Merzbach* (2000)
86 The Theory of the Design of Experiments *D.R. Cox and N. Reid* (2000)
87 Complex Stochastic Systems
O.E. Barndorff-Nielsen, D.R. Cox and C. Klüppelberg (2001)
88 Multidimensional Scaling, 2nd edition *T.F. Cox and M.A.A. Cox* (2001)
89 Algebraic Statistics—Computational Commutative Algebra in Statistics,
G. Pistone, E. Riccomagno and H.P. Wynn (2001)

Algebraic Statistics
Computational Commutative Algebra in Statistics

GIOVANNI PISTONE
EVA RICCOMAGNO
HENRY P. WYNN

CHAPMAN & HALL/CRC

Boca Raton   London   New York   Washington, D.C.

Library of Congress Cataloging-in-Publication Data

Pistone, Giovanni.
    Algebraic statistics / Giovanni Pistone, Eva Riccomagno, Henry P. Wynn.
       p. cm.-- (Monographs on statistics and applied probability ; 89)
    Includes bibliographical references and index.
    ISBN 1-58488-204-2 (alk. paper)
      1. Mathematical statistics. 2. Algebra. I. Riccomagno, Eva. II. Wynn, Henry P. III. Title. IV. Series.

QA276 .P53 2000
519.5—dc21   00-047448

This book contains information obtained from authentic and highly regarded sources. Reprinted material is quoted with permission, and sources are indicated. A wide variety of references are listed. Reasonable efforts have been made to publish reliable data and information, but the author and the publisher cannot assume responsibility for the validity of all materials or for the consequences of their use.

Neither this book nor any part may be reproduced or transmitted in any form or by any means, electronic or mechanical, including photocopying, microfilming, and recording, or by any information storage or retrieval system, without prior permission in writing from the publisher.

The consent of CRC Press LLC does not extend to copying for general distribution, for promotion, for creating new works, or for resale. Specific permission must be obtained in writing from CRC Press LLC for such copying.

Direct all inquiries to CRC Press LLC, 2000 N.W. Corporate Blvd., Boca Raton, Florida 33431.

Trademark Notice: Product or corporate names may be trademarks or registered trademarks, and are used only for identification and explanation, without intent to infringe.

© 2001 by Chapman & Hall/CRC

No claim to original U.S. Government works
International Standard Book Number 1-58488-204-2
Library of Congress Card Number 00-047448
Printed in the United States of America   1 2 3 4 5 6 7 8 9 0
Printed on acid-free paper

Contents

List of figures   ix

List of tables   xi

Preface   xiii

Notation   xv

1 Introduction   1
   1.1 Outline   1
   1.2 Computer Algebra   5
   1.3 An example: the 2^{3-1} fractional factorial design   10

2 Algebraic models   15
   2.1 Models   16
   2.2 Polynomial ideals   17
   2.3 Term-orderings   19
   2.4 Division algorithm   22
   2.5 Hilbert basis theorem   23
   2.6 Varieties and equations   25
   2.7 Gröbner bases   27
   2.8 Properties of a Gröbner basis   29
   2.9 Elimination theory   31
   2.10 Polynomial functions and quotients by ideals   33
   2.11 Hilbert function   35
   2.12 Further topics   36

3 Gröbner bases in experimental design   43
   3.1 Designs and design ideals   43
   3.2 Computing the Gröbner basis of a design   44
   3.3 Operations with designs   47
   3.4 Examples   48
   3.5 Span of a design   50
   3.6 Models and identifiability: quotients   53

3.7	Confounding of models	54
3.8	Further examples	56
3.9	The fan of an experimental design	60
3.10	Minimal and maximal fan designs	63
3.11	Hilbert functions and fans for graded ordering	65
3.12	Subsets and algorithms	66
3.13	Regression analysis	71
3.14	Non-polynomial models	72

4 Two-level factors: logic, reliability, design — 75

4.1	The binary case: Boolean representations	75
4.2	Gröbner bases and Boolean ideals	78
4.3	Logic and learning	80
4.4	Reliability: coherent systems as minimal fan designs	81
4.5	Inclusion-exclusion and tube theory	83
4.6	Two-level factorial design: contrasts and orthogonality	90

5 Probability — 95

5.1	Random variables on a finite support	96
5.2	The ring of random variables	97
5.3	Matrix representation of $\mathcal{L}(D, \mathcal{K})$	99
5.4	Uniform probability	101
5.5	Probability densities	103
5.6	Image probability and marginalisation	106
5.7	Conditional expectation	108
5.8	Algebraic representation of exponentials	111
5.9	Exponential form of a probability	113

6 Statistical modeling — 119

6.1	Introduction	119
6.2	Statistical models	120
6.3	Generating functions and exponential submodels	128
6.4	Likelihoods and sufficient statistics	131
6.5	Score function and information	135
6.6	Estimation: lattice case	136
6.7	Finitely generated cumulants	138
6.8	Estimating functions	139
6.9	An extended example	140
6.10	Orthogonality and toric ideals	149

References — 155

Index — 159

List of figures

1.1	Example of degree reverse lexicographic term-ordering	8
2.1	Example of monomial ideal	24
4.1	An input/output system	82
4.2	A simplicial complex	89
6.1	A graphical model	144

List of tables

1.1	The 2^{3-1} fractional factorial design	10
1.2	Aliasing table for the 2^{3-1} design	11
2.1	Term-orderings in three dimensions	21
2.2	Division algorithm	37
2.3	The algebra-geometry dictionary	38
2.4	Buchberger algorithm	40
4.1	Cuts and failure event	83
4.2	A fraction of the five-dimensional full factorial design	94
5.1	Values of $\mathrm{E}_0\left(X^\alpha Y^\beta\right)$ for Example 63	111
5.2	Values of $\mathrm{E}_0\left(X^\alpha Y^\beta\right)$ for Example 64	112
6.1	Z matrix of the 2^4 sample space D	141
6.2	Inverse of the Z matrix	142
6.3	Matrix $[Q(\alpha,\beta)]_{\alpha\in L;\beta=1,x_1,x_2,x_3,x_4}$	142
6.4	Matrix $[Q(\alpha,\beta)]_{\alpha\in L,\beta=x_1x_2,x_1x_3,x_1x_4,x_2x_3,x_2x_4,x_3x_4}$	143
6.5	Matrix $[Q(\alpha,\beta)]_{\alpha\in L,\beta=x_1x_2x_3,x_1x_2x_4,x_1x_3x_4,x_2x_3x_4,x_1x_2x_3x_4}$	143
6.6	Linear transformation from the θ parameters to the μ parameters	145
6.7	Matrix Z_2 for a graphical model	150

Preface

About thirty-five years ago there was an awakening of interest of researchers in commutative algebra to the algorithmic and computational aspects of their field, marked by the publication of Buckberger's thesis in 1966. His work became the starting point of a new research field, called Computational Commutative Algebra. Currently, computer programs implementing versions of his and related algorithms are readily available both as commercial products and academic prototypes. These are of growing importance in almost every field of applied mathematics because they deal with very basic problems related to systems of polynomial equations. Statisticians, too, should find many useful tools in computational commutative algebra, together with interesting and enriching new perspectives. Just as the introduction of vectors and matrices has greatly improved the mathematics of statistics, these new tools provide a further step forward by offering a constructive methodology for a basic mathematical tool in statistics and probability, that is to say a ring. The mathematical structure of real random variables is precisely a ring, and other rings and ideals appear naturally in distribution theory and modeling. However, the ring of random variables is a ring with lattice operations which are not fully incorporated into the theory we present, at least not yet.

The authors' attention was drawn to the relevance of Gröbner basis theory by a paper on contingency tables by Sturmfels and Diaconis circulated as a manuscript in 1993. With initial help provided by Professor Teo Mora (University of Genova), a first application to design of experiments was published by G. Pistone and H. Wynn in 1996 (*Biometrika*) and this field of application was more fully developed by E. Riccomagno in her Ph.D. thesis work during 1996-97 at the University of Warwick. Subsequent papers in the same direction were published by the authors and a number of coauthors. We are pleased to acknowledge (in alphabetic order) Ron Bates, Massimo Caboara, Roberto Fontana, Beatrice Giglio, Tim Holliday, Maria-Piera Rogantin.

During the few years this monograph was in the making, we have benefitted from many contributions by others, and further related work is in progress. Some of the contents of this book was first exposed at the series of four GROSTAT workshops, which took place in successive years, starting in 1997 at the University of Warwick (UK), the IUT-STID in Nice-Côte d'Azur in Menton (France), EURANDOM in Eindhoven (NL), and again, in 2000, in Menton. We must thank all the participants and these institutions for their support, in particular Professor Annie Cavarero, director of IUT-STID.

We found keen collaborators at the University of Genova. We should at least mention, together with those above, Professor Lorenzo Robbiano (who also supported GROSTAT IV) and the CoCoA team who have had a major influence on the algebraic and computational aspects of the field. We are very grateful to them all for the early and generous access to their research, for the high level of illumination it provided on the mathematical foundations and the very fast computer code developed under the wings of CoCoA.

We are grateful for many discussions with colleagues and coworkers. A minimal list includes Wilf Kendall, Thomas Richardson, Raffaella Settimi and Jim Smith, in Warwick, and Alessandro Di Bucchianico and Arjeh Cohen, in Eindhoven. Special thanks to Dan Naiman of The Johns Hopkins University for allowing us to draw on recent joint work on tube theory in Chapter 4. Ian Dinwoodie, from Tulane University, helped to strengthen our understanding of the work of Diaconis and Sturmfels on toric ideals, which we reach in the final sections of the book, from our own particular direction. Because a considerable volume of the monograph is based on work in progress, we have, on a few occasions, had to refer to unpublished, although available, technical reports. We thank all the colleagues who helped us by reading different versions of this work, some of them already mentioned, and also Neil Parkin for careful reading of the whole book. We also thank our publishers for their help and considerable patience.

A cocktail of different grants and institutions has funded this research. We should thank the UK Engineering and Physical Sciences Research Council, the Italian Consiglio Nazionale delle Richerche, EURANDOM, and, last but not least, IRMA and the University L. Pasteur of Strasbourg, and Professor Dominique Collombier, who has hosted us during the final revision of the book.

This book is dedicated to our families, with apologies to all for the absences that a triple collaboration must entail.

GIOVANNI PISTONE
EVA RICCOMAGNO
HENRY WYNN

Strasbourg, France, October 2000

Notation

Common symbols

$\mathbb{N}$	positive integer numbers
$\mathbb{Z}$	integer numbers
$\mathbb{Q}$	rational numbers
$\mathbb{R}$	real numbers
$\mathbb{C}$	complex numbers
S^*	* excludes the 0 from the set S
S_+	non-negative entries of the set of numbers S: for example $\mathbb{Z}_+ = \{a \in \mathbb{Z} : a \geq 0\} = \{0\} \cup \mathbb{N}$
d superscript	dimension of the cartesian product for example, $\mathbb{Z}^d$ stands for $\underbrace{\mathbb{Z} \times \cdots \times \mathbb{Z}}_{d \text{ times}}$
$\{a\}$	1. component-wise fractional part operator, $a \in \mathbb{R}^d$ 2. the set whose element is a
$\#A$	number of elements in the set A
$[p]$	vector or list p as a column vector
$[a_1 \cdots a_n]$	matrix with the vectors a_i, $i = 1, \ldots, n$ as columns
$[[\ldots], \ldots, [\ldots]]$	matrix as a list of rows
A^t	transpose of A where A is a matrix or a vector
I	identity matrix
$x_1, \ldots, x_d$	factors, variables, indeterminates
d	1. number of independent factors 2. number of variables 3. number of indeterminates
s	number of x_i's if the algebra is emphasised
N	1. sample size 2. number of design points 3. number of support points
$k, \mathcal{K}$	fields of coefficients for example, $\mathbb{Q}, \mathbb{R}, \mathbb{Q}(\theta)$, transcendental extension, $\mathbb{Q}(\sqrt{2})$, algebraic extension

Notation for Gröbner bases

$k[x_1, \ldots, x_s]$	ring of polynomials in $x_1, \ldots, x_s$ and with coefficients in k
$x^\alpha = x_1^{\alpha_1} \ldots x_s^{\alpha_s}$	monomial in $k[x_1, \ldots, x_s]$
$p(x_1, \ldots, x_s)$	polynomial in $k[x_1, \ldots, x_s]$
$\tau, \succ, \succ_\tau$	term-ordering
$x_{i_1} \succ \ldots \succ x_{i_s}$	initial ordering on the indeterminates
$\tau(x_{i_1} \succ \ldots \succ x_{i_s})$	emphasis on the initial ordering
$\mathrm{LT}_\tau(p(x))$	leading term of the polynomial p with respect to the term-ordering τ
$\mathrm{Ideal}(g_1, \ldots, g_h)$	ideal of $k[x_1, \ldots, x_s]$ generated by $g_1, \ldots, g_h$
$\langle g_1, \ldots, g_h \rangle$	
$\mathrm{Variety}(I)$	set of zeros of all polynomials in I
$\mathrm{Ideal}(V)$	set of all polynomials vanishing at V
$\mathrm{Variety}(f_1, \ldots, f_l)$	set of common roots of f_i, $i = 1, \ldots, l$
$\mathrm{Rem}(f), \mathrm{Rem}(f, G)$	1. normal form of f with respect to the Gröbner basis G 2. remainder of the division of f with respect to the set of polynomials G

Notation for experimental design

D, D_N	1. experimental design 2. support for a discrete distribution
a, x	design point
$x(i), (x(i)_1, \ldots, x(i)_d)$	ith design point for $i = 1, \ldots, N$
$\mathcal{X}$	design region
$\mathrm{Est}_\tau(D)$	estimable terms with respect to τ and D
$\mathcal{F}$	polynomial regression vector
$Z = [f(x)]_{x \in D, f \in \mathcal{F}}$	design matrix for a model with support $\mathcal{F}$ and a design D; the orderings on D and $\mathcal{F}$ carry over to Z
$Z^t Z$	information matrix
$y = (y_1, \ldots, y_N)$	responses, values at the support points
θ, c, b, a	parameters or coefficients
$k[x_1, \ldots, x_d]/\mathrm{Ideal}(D)$	quotient ring
$k[x]/\mathrm{Ideal}(D)$	
L	list of exponents of a vector space basis of $k[x_1, \ldots, x_d]/\mathrm{Ideal}(D)$
L_0	$L \setminus \{(0, \ldots, 0)\}$
L'	$L' \subseteq L$

Notation for logic and reliability

$\mathcal{B}(\vee, \wedge, ^-, 0, 1)$	Boolean algebra
$\vee$	maximum, union
$\wedge$	minimum, intersection
$\emptyset$	empty set
D_{2^d}	2^d full factorial design
$D \setminus D_{2^d}, \bar{D}$	complementary set of $D \subset D_{2^d}$
$f_a(x)$	polynomial indicator function of $a \in D_{2^d}$
$f_D(x)$	polynomial indicator function of $D \subset D_{2^d}$
$\mathrm{E}(f)$	mean value of f
$\triangle$	symmetric difference operator

Notation for probability and statistics

D, Ω	support of a probability space
$D^\star$	support of an image probability
A_i	elementary event
A	event
f_A	indicator function of the event A
$\mathcal{L}(D, \mathcal{K}), \mathcal{L}(D), \mathcal{L}$	the set of functions from D to $\mathcal{K}$
X	function in $\mathcal{L}(D)$
P	probability
P_0	uniform probability
K	the constant in the exponential model
$K(\Phi), K(\theta)$	cumulant generating function
$\mathrm{E}_0(X)$	expectation of X with respect to P_0
$\mathrm{E}_P(X)$	expectation of X with respect to P
m_α	raw moments $\mathrm{E}_0(X^\alpha)$
θ_α	θ-parameters of a probability
μ_α	μ-parameters $\mathrm{E}_P(X^\alpha)$
p_i	p-parameters $\mathrm{P}(a(i))$
ψ_α	ψ-parameters in exponential models
ζ_α	ζ-parameters: $\zeta_\alpha = \exp(\psi_\alpha)$
R	three-dimensional multi-array where $\mathrm{Rem}(X^{\alpha+\beta}) = \sum_{\gamma \in L} R(\alpha, \beta, \gamma) X^\gamma$
$R(\beta)$	matrix $[R(\alpha, \beta, \gamma)]_{\gamma, \alpha \in L}$
$r(\delta, \gamma)$	$R(\alpha, \beta, \gamma)$ with $\delta = \alpha + \beta$
$Q(\alpha, \beta), \alpha, \beta \in L$	$\mathrm{E}_0(X^{\alpha+\beta}) = \sum_{\gamma \in L} r(\alpha+\beta, \gamma) m_\gamma$

CHAPTER 1

Introduction

1.1 Outline

One of the most basic issues in statistical modeling is to set problems up correctly, or at least well. This means, typically, that a sample space needs to be defined together with some distribution on this sample space with some parameters. After that one can decide if the parameters or even the form of the distribution are known, and, given the motivation and resources, enter into full-blown statistical inference. Great care needs to be taken with data capture or, to put it more precisely, with experimental design, if the model is to be properly postulated, tested and used for prediction.

Some of the questions which need to be addressed in carrying out these operations are intrinsically algebraic, or can be recast as algebraic. By algebra here we will typically mean polynomial algebra. It may not at first be obvious that polynomials have a fundamental role to play.

Here is, perhaps, the simplest example possible. Suppose that two people (small enough) stand together on a bathroom scale. Our model is that the measurement is additive, so that if there is no error, and θ_1 and θ_2 are the two weights, the reading should be

$$Y = \theta_1 + \theta_2$$

Without any other information it is not possible to estimate, or compute, the individual weights θ_1 and θ_2. If there is an unknown zero correction θ_0 then $Y = \theta_0 + \theta_1 + \theta_2$ and we are in worse trouble.

In a standard regression model we write in matrix notation

$$Y = Z\theta + \varepsilon$$

and our ability to estimate the parameter vector θ, under standard theory, is equated with "Z is $N \times p$ full rank" or $\text{Rank}(Z) = p < N$ where θ is a p-vector and N is the number of design points. An example is the one-dimensional polynomial regression

$$Y(x) = \sum_{j=0}^{p-1} \theta_j x^j + \varepsilon_x$$

Then, if the experimental design consists of p distinct points $a(1), \ldots, a(p)$, the square design matrix

$$Z = \left[a(i)^j\right]_{i=1,\ldots,p;\,j=0,\ldots,p-1}$$

has full rank, and for submodels with fewer than p terms, the Z-matrix also has full rank.

Algebraic methods have been used extensively in the construction of designs with suitable properties. However, particularly in the construction of balanced factorial designs with particular aliasing properties, abstract algebra in the form of group theory has also been used to study the identifiability problem. Most students and professionals in statistics will recall a course on experimental design in which Abelian group theory is used in the form of confounding relations such as

$$I = ABC$$

and unless they are experts in experimental design, they may have remained somewhat mystified thereafter. We return to this example in Section 1.3.

Let us consider a simple example. Here is a heuristic proof that there is a unique quadratic curve through the points $(a(1), y_1), (a(2), y_2), (a(3), y_3)$

$$y_i = r(a(i)), \quad i = 1, 2, 3$$

We can think of $a(1), a(2), a(3)$ as the points of an experimental design at which we have observed y_1, y_2, y_3, respectively, without error. We also assume that $a(1), a(2), a(3)$ are distinct.

Define the polynomial

$$d(x) = (x - a(1))(x - a(2))(x - a(3))$$

whose zeros are the design points. Take any competing polynomial, $p(x)$, through the data that is such that $p(a(i)) = y_i$ (for $i = 1, 2, 3$). Write

$$p(x) = s(x)d(x) + r(x)$$

where $r(x)$ is the remainder when $p(x)$ is divided by $d(x)$. Now we can appeal to algebra and say that, given the polynomial $p(x)$, $r(x)$ is unique. But it is clear from the equation that

$$y_i = p(a(i)) = r(a(i)), \quad (i = 1, 2, 3)$$

since by construction $d(a(i)) = 0$, $i = 1, 2, 3$.

The polynomial p above can be interpreted in two ways: (i) as a continuous function with value y_i at the point $a(i)$ and (ii) as a representation of the function defined only on the design points and again with value y_i at $a(i)$ (for $i = 1, 2, 3$). The first way is very convenient when we do regression analysis and thus we call p an *interpolator*. The other interpretation is more suited for applications in discrete probability.

Here we have tried to solve an identifiability problem directly by exhibiting a minimal degree interpolator rather than check the rank of a Z-matrix.

There is a crucial point to make: *all the operations were carried out with polynomials.*

The same argument applies for polynomial regression of all orders in one dimension. However, a very important issue for this book is that if we are to use this argument for x in higher dimensions, then we need to cope with the fact that representation of points as solutions of equations, the operation of division and the remainders themselves are *not*, in general, unique in higher dimensions. The representation of discrete points as the solution of polynomial equations is to treat them as *zero-dimensional algebraic varieties*. The division operation becomes a *quotient* operation and we have jumped into algebraic geometry. The set of all polynomials which are zero on a variety (in this case, a set of points) has the algebraic structure of an *ideal*. Strictly speaking, the quotient operation uses the ideal, not the variety. The use of Gröbner bases will help throughout.

Elementary probability is not immune from this treatment. Consider a random variable X whose support is $a(1), a(2), a(3)$. What was an experimental design, above, is now a support. Since X lives only on the support, we can write (with probability one)
$$(X - a(1))(X - a(2))(X - a(3)) = 0$$
Expanding we obtain
$$X^3 = (a(1) + a(2) + a(3))X^2 - \\ (a(1)a(2) + a(1)a(3) + a(2)a(3))X + a(1)a(2)a(3)$$
Taking expectation and letting the non-central moments of X be $\mu_0 = 1$, $\mu_1 = \mathrm{E}(X)$, $\mu_2 = \mathrm{E}(X^2)$, ..., we have
$$\begin{aligned}\mu_3 &= (a(1) + a(2) + a(3))\mu_2 \\ &\quad - (a(1)a(2) + a(1)a(3) + a(2)a(3))\mu_1 \\ &\quad + a(1)a(2)a(3) \\ \mu_{3+k} &= (a(1) + a(2) + a(3))\mu_{2+k} \\ &\quad - (a(1)a(2) + a(1)a(3) + a(2)a(3))\mu_{1+k} \\ &\quad + a(1)a(2)a(3)\mu_k\end{aligned} \quad (1.1)$$

We can, in this way, express any higher-order moment as a linear function of μ_0, μ_1, μ_2. This is an example of what we shall call *moment aliasing*.

This small example points to several levels of the use of polynomial algebra in statistics. The first level is to set up the machinery for handling sets of points in many dimensions. These points will be thought of first as an experimental design D and then, when we do probability, as the support of a distribution. Of course, the problem is then different. It is the algebra which is, identical, and to emphasize this, we use the same letter D when the set of points is a support. We will cover at some length all the issues to do with description of varieties, ideals, quotient operations and so on. This occupies Chapters 2 and 3. Chapter 5 studies the algebra of random variables over a finite set of points. This is the second level.

The third level is to interpolate the probability masses for our distribution on the support D. Since the algebra has already told us how to set up interpolators, this is now straightforward, except that probabilities are non-negative and must sum to one. Still at this level we have two basic alternatives: to interpolate the raw probabilities or to interpolate their logarithm. For example, suppose we have a two-state (binary) random variable taking the values in $D = \{0, 1\}$ with probabilities $1 - q$ and q, respectively: a Bernoulli random variable. The raw interpolator is

$$p(x) = 1 - q + (2q - 1)x$$

whereas the interpolator of the logarithm, after exponentiation, gives

$$p(x) = \exp\left(\log(1 - q) + \log\left(\frac{q}{1 - q}\right) x\right)$$

The second of these is the usual exponential family representation of the Bernoulli.

The fourth level of algebraisation, and perhaps the most profound, arises from noticing that when the support D lies at integer grid points, an exponential term such as $e^{\psi_1 x_1}$ can be written $\zeta_1^{x_1}$ where

$$\zeta_1 = e^{\psi_1}$$

Using this trick, we can rewrite models in the exponential form as polynomials. For the Bernoulli, let $\psi_0 = \log(1 - q)$ and $\psi_1 = \log\left(\frac{q}{1-q}\right)$. Then, setting $\zeta_0 = e^{\psi_0}$ and $\zeta_1 = e^{\psi_1}$ we have the representation

$$p(x) = \zeta_0 \zeta_1^x$$

This coincides with the familiar form $p(x) = q^x (1 - q)^{1-x}$. We shall also discuss this form, which is closely related to the work of Diaconis and Sturmfels (1998) on toric ideals.

Note that we have been a little lazy with the notation here. All the forms of $p(x)$ have a different structure but agree numerically on D.

Much of the real usefulness of algebra in statistics comes from the interplay between these different parametrisations. We shall also need another parametrisation in terms of moments. This is made harder by the fact that, typically, statistical models or submodels are obtained by imposing restrictions on the parameters. We shall define an *algebraic statistical model* as one which adopts one of these representations and for which the restrictions on the parameters are themselves polynomial. However, and this is the most complex issue in the book, the forms of these submodels may be different depending on the parametrisation. Only sometimes can they be perfectly linked. An important example is the independence condition, which forces factorisation of the raw polynomial interpolators, maps to additivity inside the exponential representation and factorisation in the ζ and q^x forms. Conditional independence, as used in Bayes networks, also has this multiple representation. Chapters 5 and 6 discuss all these issues.

The book can be seen from different angles and we are grateful to a reviewer for making us more aware of this. The ambitious angle, and more relevant to researchers in statistics, is to rewrite the foundations of discrete probability and statistics in the language of algebraic geometry. We have only partly succeeded in doing this. There is still much to be done, particularly in sorting out fundamental issues arising from submodels discussed in the last chapter, both theoretically and computationally. This effort must surely draw on the important work of Andrews and Stafford (2000) on general application of computer algebra to statistics.

The more modest objective in which we hope to have succeeded is to enlarge the kitbag of tools available to the statistician. The Gröbner basis method in experimental design can now be used routinely, and is by the authors, to investigate the identifiability of experimental design/model combinations in real applications. The use of the methods in statistical modeling should also proceed rapidly. After the seminal work by Diaconis and Sturmfels (1998), there have been advances in using Gröbner basis methods for Monte Carlo style sampling on contingency tables, notably by Dinwoodie (1998). Promising ongoing work on the use of Gröbner basis methods in Bayes networks is being carried out by J. Q. Smith and R. Settimi. We also include in Section 4.5 work by the authors and other collaborators on reliability on binary (two-level) factorial design.

1.2 Computer Algebra

Several packages for symbolic computation and Gröbner basis computation are available: CoCoA, Maple, Mathematica and GB, to mention a few. We have used mostly Maple and CoCoA. Some points need to be made about these packages.

The package CoCoA (COmputations in COmmutative Algebra, freely available at http://cocoa.dima.unige.it) is specially developed for research in algebraic geometry and commutative algebra. Thus it is faster than most other software in computing Gröbner bases, although at times not intuitive, and it allows more refined computations. The interface needs further development and the use of unknown constants is not implemented. Nevertheless in some cases ad hoc tricks can be used to force some indeterminates to play the role of unknown constants. An example is the case of complex numbers for which an indeterminate i is introduced to represent the complex unit. For details see Caboara and Riccomagno (1998).

Robbiano and other members of the CoCoA team are very active in the research area described in Chapters 2 and 3 of this book. They concentrate mainly on links to algebraic geometry with forays into statistics (Robbiano and Rogantin (1998), Caboara and Robbiano (1997)), while the authors are led by applications in statistics with some expeditions into the mathematics and computation.

Maple (University of Waterloo, Canada http://www.maplesoft.com) is a general purpose package for symbolic computations. It is quite fast, simple to use and with a good online help. It has a very good interface, allows the use of unknown constants or free parameters, but it is slower than CoCoA for the specialized application described here. Maple V-5 includes the package Groebner for doing Gröbner basis computation, and allows the use of unknown constants and user-defined term-orderings.

Sometimes our examples will be over the set of integers, $\mathbb{Z}$, which is not a field. Gröbner basis theory has a counterpart for polynomials with integer coefficients, but it is more expensive. For example, in CoCoA, when the ring $\mathbb{Z}[x_1, x_2]$ is input, a message appears warning that *G-basis-related computations could fail to terminate or can be wrong*. However, $\mathbb{Z}$ is embedded in $\mathbb{Q}$, and one can work with rational coefficients and multiply everything out to obtain integers. On other occasions one has to work with a finite set of coefficients, say $\mathbb{Z}_p$. For p, a prime integer, $\mathbb{Z}_p$ forms a field and the algebraic theory of Gröbner bases is similar to that over rational numbers. In other cases, such as the trigonometric case (see Section 3.14), difficulties arise from the fact that the sine and cosine of rational values are typically irrational numbers and thus the coefficient field is not embeddable in $\mathbb{Q}$. Ad hoc procedures have been considered based on simple algebraic extensions of rational numbers.

As mentioned, the authors prefer to use Maple and CoCoA. Lists of software that include routines to compute Gröbner bases are maintained at http://SymbolicNet.mcs.kent.edu/ and http://anillos.ugr.es/. We should mention: Matematica for its popularity, REDUCE written in LISP and whose main characteristics are code stability, full source code availability and portability, and AXIOM, which takes an object-oriented approach to computer algebra and its overall structure is strongly typed and hierarchical. Among the freely available software there is GROEBNER (at ftp.risc.uni-linz.ac.at) developed at RISC-Linz by W. Windsteiger and B. Buchberger, Macaulay2 (http://www.math.uiuc.edu/Macaulay2/) developed by D. Grayson and M. Stillman to support research in algebraic geometry and in commutative algebra. The package SINGULAR (http://www.singular.uni-kl.de/) is advertised as the most powerful and efficient systems for polynomial computations with a kernel written in C++.

Next we anticipate some notions from Chapter 2. Historically a first application of Gröbner bases is as polynomial system solver in that it can rewrite a system of polynomial equations in an equivalent form which is easier to solve. Equivalent means with the same set of solutions. For example, if the system has a finite number of solutions, there is a Gröbner basis including a polynomial in only one indeterminate, a polynomial in that indeterminate and another one, and so on. In this way the system can be solved by backward substitution. The great advantage of Gröbner bases with respect to, say, numerical methods for solving systems of polynomial equations, is that it can also be used when the system has infinitely many solutions. All the solutions are returned but in a parametric, or implicit, form, which sometimes seems even more complicated than the original. This is why it is generally recommended to couple Gröbner basis with numerical methods when used as system solver.

In this book we are concerned with two slightly different algebraic aspects which use the same Gröbner basis techniques. 1. We know the solutions (so to speak) and are interested in determining the set of polynomials interpolating them. Then, Gröbner basis methods return a basis of the set of functions defined over the solutions. 2. We have a system of polynomial equations and would like to check whether there are some algebraic relations. That is, we need to rewrite the system in a different form. The operations we allow are sums of elements in the polynomial set considered and products with any polynomial. This leads to the definition of a polynomial ideal for which we refer to the main text.

1.2.1 A quick introduction to Gröbner bases

A polynomial, in one or more variables, is a linear combination of monomials. Thus $1 + 2x_1 + 3x_2 + 4x_1x_2$ is a polynomial and $1, x_1, x_2, x_1x_2$ are monomials.

On the set of integer numbers there is one natural total order, the one we all know. The set of monomials in one indeterminate, x, inherits such an order, thus x is lower than x^3 and $1 = x^0$ is lower than x^α for all α positive integers. We do not consider negative integers.

In more than one dimension the uniqueness of a natural way of ordering points on the (non-negative) integer grid is lost. The same is valid for monomials in more than one indeterminate. In Chapter 2 monomial orderings (also called term-orderings) are properly defined. For the moment we only observe that a term-ordering corresponds to a total order on the integer grid and is compatible with cancellation of monomials. There are orderings on the integer grid that do not correspond to any term-ordering.

The most common term-ordering is the lexicographic ordering. In three dimensions x, y and z, first fix z larger (in the ordering) than y and y larger than x. We write $z \succ y \succ x$ and talk of initial ordering. All monomials of the type x^α are lower than any monomial involving y and/or z and the monomials x^α are ordered according to the one-dimensional ordering. Next come the monomials with the y indeterminate at first degree, that is $x^\alpha y$, which are again ordered according to the one-dimensional ordering. After

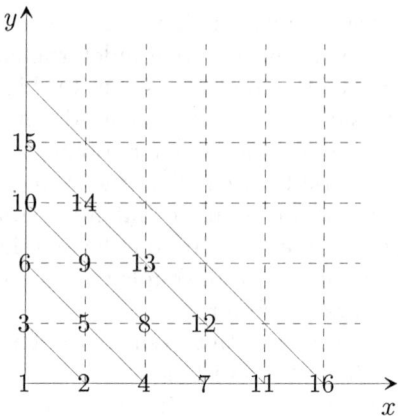

Figure 1.1 *Example of degree reverse lexicographic term-ordering in two dimensions.*

that we have the monomials $x^\alpha y^2$. After all the monomials $x^\alpha y^\beta$, for α and β non-negative integers, it is the turn of the monomials including the z indeterminate.

The degree reverse lexicographic term-ordering is a term-ordering often used. An example in two dimensions is given in Figure 1.1. Monomials on a line parallel to $y = -x$ are ordered in a linear fashion according to the ordering in one dimension and going in the direction bottom to top, that is x^α is smaller than y^α. Monomials on lines closer to the origin are smaller than monomials on lines far away. In higher dimensions, hyper-planes play the role of lines. For a definition see Section 2.3.

Once a term-ordering is chosen, the largest term of a polynomial is well defined and is called its *leading term*.

Consider the system of polynomials

$$\begin{cases} yx - z \\ x^2 - z \end{cases} \quad (1.2)$$

The associated system of equations is obtained by equating to zero the two polynomials. A quick computation shows that there are two sets of solutions

$$\begin{cases} x = 0 \\ y = y \\ z = 0 \end{cases} \quad \text{and} \quad \begin{cases} x = y \\ y = y \\ z = y^2 \end{cases}$$

The following systems of polynomial equations have the same solutions, that is they are algebraically equivalent,

$$\begin{cases} (y-x)x = 0 \\ z - x^2 = 0 \end{cases} \qquad \begin{cases} yx - z = 0 \\ z(y-x) = 0 \\ z - x^2 = 0 \end{cases}$$

The corresponding two sets of polynomials are two different Gröbner bases of the ideal generated by Equation (1.2) with respect to two different term-orderings. That is the lexicographic ordering with initial ordering $z \succ y \succ x$ and the degree reverse lexicographic term-ordering with the same initial ordering, respectively. The leading terms are $\{yx, z\}$ and $\{yx, zy, x^2\}$.

Looking at the solutions of the systems, one is tempted to say that an equivalent set of polynomials is

$$\begin{cases} x - y \\ z - y^2 \end{cases} \qquad (1.3)$$

But it cannot be retrieved from the polynomials in (1.2) using sums and products of polynomials. That is, this last system is not algebraically equivalent to the others. The solution $(0,0,0)$ is clearly given in (1.2) while in (1.3) it is deduced from the solution $x = y$, $z = y^2$ for $y = 0$. This phenomenon is referred to as the multiplicity of a solution.

Roughly speaking, Gröbner basis computation allows us to rewrite the system (1.2) without losing or adding solutions, by having the correct set of leading terms. Namely, a polynomial set G is a Gröbner basis for a set of polynomials F and with respect to a term-ordering if the set of polynomials generated by the leading terms of F is equal to the analogous set generated by the leading terms of G. The elements of the set generated by the polynomials $\{f_1, \ldots, f_s\}$ are the polynomials $\sum_{i=1}^{s} h_i f_i$, where the h_i's are generic polynomials. Note the role of a term-ordering in the definition of Gröbner bases. The set of polynomials $F = \{f_1 = yx - z, f_2 = x^2 y - z\}$ does not form a Gröbner basis with respect to the lexicographic term-ordering with initial ordering $z \succ y \succ x$. Call this term-ordering τ. Indeed yx cannot be obtained from the leading terms of f_1 and f_2, which is z for both f_1 and f_2, but it is the leading term of $f_1 + f_2$. The (reduced) Gröbner basis of F with respect to τ is given above. There is an algorithm to compute Gröbner bases given a set of polynomials and a term-ordering which is described in Section 2.12.3.

Having the right leading terms also helps in the division of polynomials. Namely the division of a polynomial by a Gröbner basis has a unique remainder, while in general this is not true. The division of a polynomial f by a set F is a way of rewriting f as a polynomial combination of elements of F in such a way that we are left with a reminder whose leading term is not divisible by the leading terms of the polynomials in F. For example consider $f = z$. The division of f by f_1 and f_2, with respect to τ, gives the reminder yx if we divide first by f_1, indeed $f = (-1)f_1 + xy$. But if we first

Table 1.1 *The 2^{3-1} fractional factorial design.*

A	B	C
1	1	1
1	−1	−1
−1	1	−1
−1	−1	1

divide by f_2 and then by f_1 we obtain $f = (-1)f_2 + x^2$ where now x^2 is the remainder. Fortunately when we divide f with respect to the Gröbner basis G, we do not need to consider with respect to which polynomial we divide first, the reminder will always be the same, z itself in this example.

1.3 An example: the 2^{3-1} fractional factorial design

In this section we outline the ideas and techniques presented in this book on an example which we shall return to in the main text as well. Consider the four points of the 2^{3-1} fractional factorial design with levels ±1 in Table 1.1 (see Box, Hunter and Hunter (1978) and Cox and Reid (2000)). It is defined by the confounding relation $ABC = I$ where A, B and C are the factors and I is the identity. When we refer to the factors in the classical framework, for example when using the mathematics of group theory, we use capital letters. We use small letters a, b and c for factors in our polynomial representation. Moreover some computer algebra software require that indeterminates, the algebraic equivalent of factors, are a single, small letter.

The rows in Table 1.1 are solutions of the following system of polynomial equations, which defines the 2^{3-1} design

$$\begin{cases} a^2 - 1 = 0 \\ b^2 - 1 = 0 \\ c^2 - 1 = 0 \\ abc - 1 = 0 \end{cases} \quad (1.4)$$

The aliasing table in Table 1.2 is obtained by multiplying $ABC = I$ by A, B and C, respectively. Now, the system of polynomial equations originated by substituting small letters in Table 1.2 has still the same set of solutions as the system in (1.4). For the polynomials in the system so obtained, namely $abc - 1$, $bc - a$, $ac - b$, $ab - c$, the first polynomial is larger than the other three polynomials as its highest term is divided by the second-order terms of the other three polynomials. In this sense it is redundant

AN EXAMPLE: THE 2^{3-1} FRACTIONAL FACTORIAL DESIGN

Table 1.2 *Aliasing table for the 2^{3-1} design.*

$$\begin{aligned} ABC &= I \\ BC &= A \\ AC &= B \\ AB &= C \end{aligned}$$

and it can be substituted by the three polynomials $a^2 - 1$, $b^2 - 1$ and $c^2 - 1$ which are of smaller order. The set of zeros of the system of polynomial equations obtained equating to zero these new three polynomials is the 2^3 full factorial design.

The final set of equations so obtained forms a Gröbner basis

$$\begin{cases} a^2 - 1 \\ b^2 - 1 \\ c^2 - 1 \\ bc - a \\ ac - b \\ ab - c \end{cases} \tag{1.5}$$

General methods to compute Gröbner bases from a set of polynomials are given in Chapter 2.

In the classical theory, one would look at the aliasing table in Table 1.2 and deduce that the interaction AB is aliased to the linear factor C. That is the effects of AB and C are confounded and both AB and C cannot be terms in the same linear regression model. In more mathematical terminology one says that AB and C are linearly dependent functions over the 2^{3-1} design. The approach presented in this book develops this observation. The theory of Gröbner basis automatises the process of finding a vector space basis of the set of functions defined over the 2^{3-1} design. From this vector space basis it is easy to check whether two terms are confounded. This saturated set of independent terms is formed by monomials, that is factors and interactions. It will be the basis with the terms smallest in some sense which will be clear when in Chapter 2 the concept of term-ordering is explained.

We show the process for determining this vector space basis for the 2^{3-1} design. Consider the Gröbner basis in Equation (1.5) and consider the largest terms of each of its polynomials, they are

$$\text{LT} = \{a^2, \ b^2, \ c^2, \ ab, \ ac, \ bc\}$$

The formalization of this process requires again the definition of term-ordering. For the moment it is sufficient to say that, for example, in $ab - c$

the term ab is larger than c because it represents a second-order interaction. In some cases to be considered later it will be possible that a linear term is larger than an interaction.

Now consider all the terms that are not divided by the monomials in LT. They are listed below and they are four, exactly the number of points in the 2^{3-1} design:

$$1, \quad a, \quad b, \quad c$$

The theory of Gröbner bases states that this is a set of linearly independent functions over the 2^{3-1} design. They can be used to build a linear regression model.

In particular all the functions over the 2^{3-1} design can be represented as linear combinations of those four monomials, and a function f is written as

$$f(x) = \theta_0 + \theta_1 a + \theta_2 b + \theta_3 c$$

where x ranges over the points in the 2^{3-1} design. Now probabilities are functions and thus can be represented in this way, and the θ coefficients are chosen so that $\sum_{x \in 2^{3-1}} f(x) = 1$. For example, the probability that assigns mass $1/2$ to the point $(1,1,1)$, mass $1/4$ to the point $(-1,1,-1)$, and equal mass $1/8$ to the other two points is the function

$$1/4 + 1/16a + 1/8b + 1/16c$$

The uniform probability is given by the constant function $1/4$.

Random variables are again linear functions of $1, a, b, c$, for example $Y = A + B + C$. The expectation of Y with respect to the uniform probability can now be computed with linear operations as

$$\mathrm{E}_0(Y) = \sum_{x \in 2^{3-1}} Y(x) = \sum_{(a,b,c) \in 2^{3-1}} (a + b + c) = 0$$

Analogously, the second-order moment is

$$\mathrm{E}_0(Y^2) = \sum_{x \in 2^{3-1}} Y(x)^2 = \sum_{(a,b,c) \in 2^{3-1}} (a + b + c)^2 = 12$$

As mentioned previously the relation (1.1) further simplifies the computation of higher-order moments.

We conclude this section by computing the image probability of Y. Let us start with the computation of the image support. Thus adjoin the polynomial for Y, using small letter y, to the equations of the Gröbner basis of

AN EXAMPLE: THE 2^{3-1} FRACTIONAL FACTORIAL DESIGN

the 2^{3-1} design

$$\begin{cases} a^2 - 1 \\ b^2 - 1 \\ c^2 - 1 \\ bc - a \\ ac - b \\ ab - c \\ y - (a + b + c) \end{cases} \quad (1.6)$$

The aim is to find a polynomial involving only y and not the indeterminates a, b and c. That is to check whether y is algebraically independent from a, b and c. The square of the last polynomial in (1.6) above gives

$$y^2 + (a + b + c)^2 - 2y(a + b + c)$$

and thus, using again the definition of y,

$$y^2 - (a + b + c)^2 = y^2 - a^2 - b^2 - c^2 - 2bc - 2ac - 2ab$$

Now $a^2 = b^2 = c^2 = 1$ and $bc = a$, $ac = b$ and $ab = c$, giving

$$y^2 - 2y - 3 = (y + 1)(y - 3) = 0$$

This is the description of the image of Y. In Chapter 5 this process is automatised by considering the Gröbner basis of the polynomials above with respect to a so-called elimination term-ordering.

The image probability of Y takes values on the set $D^* = \{-1, 3\}$ and its density with respect to the uniform distribution has the form of a polynomial supported on $\{1, y\}$. Thus in generic form we can write

$$p_Y = \theta_0 + \theta_1 Y$$

Because the support of p_Y is $\{1, y\}$, the density p_Y is fully known if the first two moments $\mathrm{E}(Y^\alpha)$, $\alpha = 0, 1$ are known. By using the conditions $Y^2 = 2Y + 3$, $\mathrm{E}(Y) = 0$, and $\mathrm{E}_*(Y) = \dfrac{-1 + 3}{2} = 1$ (the expectation with respect to the uniform on D^*), we obtain the system

$$\begin{cases} 1 = \mathrm{E}_*(\theta_0 + \theta_1 Y) = \theta_0 + \theta_1 \\ 0 = \mathrm{E}_*(\theta_0 Y + \theta_1 Y^2) = \mathrm{E}_*(\theta_0 Y + \theta_1(2Y + 3)) = \theta_0 + 5\theta_1 \end{cases}$$

which gives $p_Y = \dfrac{5}{4} - \dfrac{1}{4}Y$.

The polynomial setup presented here can be used to discuss many probabilistic and statistical concepts. Much of this can be found in the main text but still much work is left for the authors and the interested reader.

CHAPTER 2
Algebraic models

In this chapter we introduce the basic algebraic machinery: rings of polynomials, ideals, varieties and Gröbner bases. The effort is motivated by a general definition of a model to be given in Section 2.1.

In our definition of model, factors or inputs are denoted by the letter x, responses or outputs are denoted by y, parametric functions denoted by θ or functions of θ. These are related by polynomial algebraic relations, possibly implicit. Another feature of the definition is that constraints of polynomial type may be included in the specification of the model. Implicit models and the introduction of constraints can lead to the use of dummy variables. All this requires complex polynomial computations to be tackled with advanced tools from polynomial ring theory together with their computer implementation.

In this algebraic framework, the parameters of the model as interpreted in statistics are functions of any form with the restriction that they belong to a specified field. For example, $\mathbb{Q}(\theta_1, \ldots, \theta_p)$ is the set of all rational functions in $\theta_1, \ldots, \theta_p$ with rational coefficients. Another example is $\mathbb{Q}(e^{\theta_1}, \ldots, e^{\theta_p})$, the set of all exponential rational functions. Parameters are treated as unknown quantities and in most of the cases they appear in linear form.

The process of actual estimation of parameter values, given observed values of factors and responses, will be formalised in Chapter 6. For the purpose of the present chapter it suffices to represent the statistical error (deviation from the model) by dummy variables.

Our working algebraic space is $k[x_1, \ldots, x_s]$, the commutative ring of all polynomials in the indeterminates $x_1, \ldots, x_s$ and with coefficients in k, where k is a field. Most of the time k will be the rational numbers $\mathbb{Q}$, or a field extension of $\mathbb{Q}$.

There is often no ordering on the indeterminates $x_1, \ldots, x_s$, in particular no ordering is necessary when talking of ideals in $k[x_1, \ldots, x_s]$ and varieties in k^s (see Definition 5). Nevertheless in some algebraic and statistical situations it is necessary to order indeterminates.

Definition 1 *An* initial ordering *is a total order on the indeterminates* $x_1, \ldots, x_s$.

When the indeterminates are indexed from 1 to s, such as $x_1, \ldots, x_s$, it is convention to consider an initial ordering $x_i \succ x_{i+1}$ for all $i = 1, \ldots, s-1$.

In certain cases specific knowledge of the problem intervenes in the choice of the initial ordering. In Section 3.9 we use the idea of the fan of an ideal to drop the assumption of having an *a priori* initial ordering (and in general a term-ordering, see Section 2.3).

The quantities of the form $x_1^{\alpha_1} \ldots x_s^{\alpha_s}$ with $\alpha_i \in \mathbb{Z}_+$ for all $i = 1, \ldots, s$ are called *terms*.

Definition 2 *The set of all terms in s indeterminates is denoted by T^s.*

Note that given an initial ordering a term is specified by the vector of length s of its exponents. And thus T^s is coded by $\mathbb{Z}_+^s$. Specifically this is the multi-index representation or log function of Definition 3.

Definition 3 *Let $x_1, \ldots, x_s$ be indeterminates and let the initial ordering be $x_i \succ x_{i+1}$ for all $i = 1, \ldots, s-1$. The log operator is the function*

$$\log : \quad T^s \quad \longrightarrow \quad \mathbb{Z}_+^s$$
$$x^\alpha = (x_1^{\alpha_1}, \ldots, x_s^{\alpha_s}) \quad \longmapsto \quad (\alpha_1, \ldots, \alpha_s)$$

For example in $\mathbb{Q}[a, b, c]$ with the initial ordering $a \succ b \succ c$ we represent the term $a^7 b^2$ as $\log(a^7 b^2) = (7, 2, 0)$ and with the initial ordering $b \succ c \succ a$ as $\log(a^7 b^2) = (2, 0, 7)$.

Note that an ordering over $\mathbb{Z}_+^s$ translates via logarithm into an ordering over T^s. An ordering over T^s which is compatible with monomial cancellation is called a term-ordering. Term-orderings are discussed in detail in Section 2.3 below.

The elements of $k[x_1, \ldots, x_s]$ of the form $a x_1^{\alpha_1} \ldots x_s^{\alpha_s}$ with $\alpha_i \in \mathbb{Z}_+$ for all $i = 1, \ldots, s$ and a in k are called monomials. The term $x_1^{\alpha_1} \ldots x_s^{\alpha_s}$ is identified with the monomial $a x_1^{\alpha_1} \ldots x_s^{\alpha_s}$ for $a = 1$. Finally a polynomial is a k-linear combination of terms.

A polynomial function is associated to each polynomial f as follows:

$$f : \quad k^s \quad \longrightarrow \quad k$$
$$(a_1, \ldots, a_s) \quad \longmapsto \quad f(a_1, \ldots, a_s)$$

In our definition a model is described as a set of polynomial equations. The algebraic variety of the finite set of polynomials $f_1, \ldots, f_r$ in $k[x_1, \ldots, x_s]$ is the set

$$\text{Variety}(f_1, \ldots, f_r) = \{(a_1, \ldots, a_s) \in k^s : f_j(a_1, \ldots, a_s) = 0, \; j = 1, \ldots, r\}$$

Note that the variety depends on the specified field k. That is, the set of solutions of the system of equations defining the variety depends on the space in which the computation is done. For example, the univariate polynomial $x^2 + 1$ does not have solutions over the real numbers (empty variety) and has two distinct solutions over the complex number field, namely $\pm i$.

2.1 Models

Definition 4 *Let k be a field, called the field of constants. Let $\mathcal{K}$ be a field of functions $\phi : \Theta \longrightarrow k$, with Θ the set of parameters; $\mathcal{K}$ is called the* field of parametric functions. Let $x = (x_1, \ldots, x_d)$ be the control factors, $y = (y_1, \ldots, y_p)$ be the responses and $t = (t_1, \ldots, t_h)$ be dummy variables. A model is a finite list of polynomials, $f_1, \ldots, f_q, h_1, \ldots, h_l$, such that $f_i \in \mathcal{K}[x, y, t]$ and $h_j \in k[x, t]$. The variety Variety$(f_i, h_j : i = 1, \ldots, q; j = 1, \ldots, l) \in \mathcal{K}^{d+p+h}$ is called the model variety and the variety Variety$(h_j) \in k^{d+h}$ is called the input variety.

Example 1 The linear model $y = a + bx$ is $f_1 = y - a - bx \in \mathbb{Q}(a, b)[x, y]$.

Example 2 The model $y = ax_1 + bx_2 + cx_3$ over the mixture constraint $x_1 + x_2 + x_3 = 1$ becomes $f_1 = y - ax_1 - bx_2 - cx_3, h_1 = x_1 + x_2 + x_3 - 1 \in \mathbb{Q}(a, b, c)[x_1, x_2, x_3, y]$. Another example of a mixture model is $f_1 = \theta_1 x_1^2 + \theta_2 x_2^2 + \theta_3 x_3$ with the additional constraint $x_1 + x_2 + x_3 - 1$ in $\mathbb{Q}(\theta_1, \theta_2, \theta_3)[x_1, x_2, x_3]$.

Example 3 The implicit model $y = \dfrac{1}{a + bx}$ can be written in $\mathbb{Q}(a, b)[x, y]$ as $f_1 = (a + bx)y - 1$. The condition that $a + bx \neq 0$ is included by adjoining a dummy variable, $f_1 = (a + bx)y - 1, f_2 = (a + bx)t - 1 \in \mathbb{Q}(a, b)[x, y, t]$.

Example 4 The inverse polynomial model, see McCullagh and Nelder (1983), is $x/y = a + bx$. We write $f_1 = bxy + ay - x \in \mathbb{Q}(a, b)[x, y]$. To represent an unspecified shift of x, a constant c is introduced by writing $f_1 = bty + ay - t, f_2 = t - x + c$ in $\mathbb{Q}(a, b, c)[x, y, t]$.

Example 5 In this model we observe the distance from the origin of a circle with unknown centre (a, b) and radius r in the directions x_1, x_2. The model in $\mathbb{Q}(a, b, r)[x_1, x_2, y, t_1, t_2]$ is

$$f_1 = (t_1 - a)^2 + (t_2 - b)^2 - r^2$$
$$f_2 = t_1 x_2 - t_2 x_1$$
$$f_3 = y^2 - t_1^2 - t_2^2$$
$$h_1 = x_1^2 + x_2^2 - 1$$

where f_3 gives the observed values, h_1 the unit circle, f_1 a generic circle and f_2 the mapping between the circles in f_1 and h_1. This model is implicit and, in contrast to Example 4, the equations do not define a unique value of y given the parameters and the indeterminates.

Example 6 In this example the statistical error is represented by dummy variables. Let $t = (\eta, \varepsilon)$ and $f_1 = \eta - a - bx$, $f_2 = y - \eta - \varepsilon$, $h_1 = x(x - 1)(x - 2)$. The polynomial f_1 models the mean response $\eta = a + bx$, f_2 the deviation from the mean $\varepsilon = y - \eta$ and h_1 defines the design points $\{0, 1, 2\}$.

2.2 Polynomial ideals

A basic tool is the algebraic structure called an ideal. A polynomial ideal formalises the intuitive idea of the algebraic consequences of a system of polynomial equations. Following the notation generally used in algebra, in the remaining sections of this chapter we indicate indeterminates by $x_1, \ldots, x_s$ or in short x.

Definition 5

1. A (polynomial) ideal I is a subset of a polynomial ring $k[x]$ closed under sum and product by elements of $k[x]$. Specifically the set $I \subset k[x]$ is an ideal if for all $f, g \in I$ and $h \in k[x]$ the polynomials $f + g$ and hf are in I.

2. Let F be a set of polynomials. The ideal generated by F is the smallest ideal containing F. It is denoted $\langle F \rangle$.

3. An ideal I is radical if $f \in I$ whenever a positive integer m exists such that $f^m \in I$.

4. The radical of an ideal is the radical ideal defined as

$$\sqrt{I} = \{f \in k[x] : \text{ a positive integer } m \text{ exists such that } f^m \in I\}.$$

Example 7 The set of all one-dimensional polynomials of the form $xp(x)$, where p ranges over the polynomial ring is a radical ideal. In fact if f^m has a zero constant term, then f has a zero constant term. Notice that the set $\{x^2 p(x)\}$, p polynomial, is not a radical ideal.

Another example of an ideal is included in the following definition.

Definition 6 The projections of the ideal I with respect to a subset of indeterminates S is the ideal $I \cap k[S]$. In particular $I_p = I \cap k[x_{p+1}, \ldots, x_s]$ is called the p-th elimination ideal of I.

Elimination ideals are extensively used in *elimination theory* (see Section 2.9), for example to triangularise a system of polynomial equations and in other important applications.

Let us consider a finite set F of polynomials. All the polynomials generated via ideal operations form the smallest ideal containing F, that is the ideal generated by F.

Definition 7 An ideal I is finitely generated if there exist $f_1, \ldots, f_r$ polynomials in $k[x]$ such that for any $f \in I$ there exist $s_1, \ldots, s_r$ polynomials of $k[x]$ such that

$$f = \sum_{i=1}^{r} s_i f_i$$

We write $I = \langle f_1, \ldots, f_r \rangle$ and the set $\{f_1, \ldots, f_r\}$ is called a basis of I.

The Hilbert basis theorem (see Section 2.5) states that any polynomial ideal is finitely generated.

We call $f = \sum_{i=1}^{r} h_i f_i$ a polynomial combination by analogy with linear combinations of vectors. In contrast to vector spaces where the elements of a basis must span and be linearly independent over k, an ideal basis need only span. For example, in $I = \langle x_1, x_2 \rangle \subset k[x_1, x_2]$ the polynomial $f = 0$ can be expressed both as $f = 0 \cdot x_1 + 0 \cdot x_2$ and $f = x_2 \cdot x_1 - x_1 \cdot x_2$. In more than one-dimension this lack of independence, due to the fact that the coefficients are polynomials, is a cause of difficulties, for example in extending the standard polynomial division algorithm for one dimension. Moreover, a polynomial in an ideal may be expressed as a polynomial combination of the basis elements in different but equivalent ways. Two bases of the same ideal are two different but equivalent ways to write the same set of polynomial equations and it is possible to write one in terms of the other with ideal operations. Representations over different bases will be shown to have different statistical interpretations, so that the ability to move between different representations becomes extremely interesting.

In the last example we note that $\{x_1, x_2\}$ is a *minimal* basis for I, that is none of its proper subsets is a basis for I. Minimal bases of the same ideal can consist of different numbers of elements. For example, we have $\langle x^2, x + x^2 \rangle = \langle x \rangle$ and both are minimal. In one dimension there is a privileged generator of an ideal $\langle f_1(x), \ldots, f_v(x) \rangle$ consisting of a single element, namely the greatest common divisor or GCD of $f_1, \ldots, f_v$.

2.3 Term-orderings

Univariate polynomials are linear combinations of univariate terms x^n, which are ordered by their degree. All computations for one-dimensional polynomials exploit this fact. In more than one dimension it is necessary to introduce the concept of a term-ordering to order terms. Term-orderings express a ranking of the algebraic complexity of a model structure.

Note that the terms of T^s are naturally pre-ordered according to simplification of terms. For example $x_1^2 x_3$ precedes $x_1^3 x_3^2$ as the "fraction" $\frac{x_1^3 x_3^2}{x_1^2 x_3} = x_1 x_3$ is still in T^s.

Definition 8 *A monomial or term-ordering on $k[x]$ is an ordering relation $\succ_\tau$ (or τ or $\succ$) on T^s, that is the terms of $k[x]$, satisfying*

1. $x^\alpha \succ 1$ for all x^α with $\alpha \neq 0$ and

2. for all $\alpha, \beta, \gamma \in \mathbb{Z}_+^s$ such that $x^\alpha \succ x^\beta$, then $x^\alpha x^\gamma \succ x^\beta x^\gamma$.

We will also use the notation $x^\beta \prec x^\alpha$ for $x^\alpha \succ x^\beta$. Note that the restriction of a term-ordering to the terms of the type x_i gives an initial ordering of the indeterminates of $k[x_1, \ldots, x_s]$. Definition 8 implies the following.

(i) Any two terms are comparable, that is for any x^α, x^β either $x^\alpha \succ x^\beta$ or $x^\alpha = x^\beta$ or $x^\beta \succ x^\alpha$. This property characterises total orderings.

(ii) There is no infinite descending chain, that is any subset of terms contains a minimum element with respect to the ordering. This property is known as well ordering.

(iii) The ordering is compatible with the simplification of terms, that is for any pair of terms x^α and x^β, if x^α divides x^β then $x^\beta \succ x^\alpha$.

There are two basic term-orderings.

Example 8 The lexicographic term-ordering for which $x^\alpha \succ_{\texttt{lex}} x^\beta$ if with reference to the log representation in the vector $\alpha - \beta$ the left-most nonzero entry is positive. That is

$$x^\alpha \succ x^\beta \quad \text{if and only if} \quad \begin{cases} \alpha_1 > \beta_1 \\ \text{or there exists } p \leq s \text{ such that} \\ \alpha_i = \beta_i \text{ for } i = 1, \ldots, p-1 \text{ and } \alpha_p > \beta_p \end{cases}$$

For example for $x_1 \succ x_2$, $x_1{}^2 x_2{}^5 \succ_{\texttt{lex}} x_1 x_2^{12}$ and $x_1^2 x_2^2 \succ_{\texttt{lex}} x_1^2 x_2$. In a lexicographic ordering an indeterminate dominates over the others.

Example 9 The other basic term-ordering is $\texttt{tdeg}$ or degree reverse lexicographic term-ordering, for which $x^\alpha \succ_{\texttt{tdeg}} x^\beta$ if $\sum_{i=1}^s \alpha_i > \sum_{i=1}^s \beta_i$ or $\sum_{i=1}^s \alpha_i = \sum_{i=1}^s \beta_i$ and the right-most nonzero entry of $\alpha - \beta$ is negative. That is

$$x^\alpha \succ x^\beta \quad \text{if and only if} \quad \begin{cases} \sum_{i=1}^s \alpha_i > \sum_{i=1}^s \beta_i \\ \text{or there exists } p \leq s \text{ such that} \\ \alpha_i = \beta_i \text{ for } i = p+1, \ldots, s \text{ and } \beta_p > \alpha_p \end{cases}$$

For example $x_1 x_2^5 \succ_{\texttt{tdeg}} x_1^2 x_2^3$, $x_1^2 \succ_{\texttt{tdeg}} x_1 x_2$, assuming that $x_1 \succ x_2$. In the $\texttt{tdeg}$ ordering terms are ordered according to their total degree $\sum_{i=1}^s \alpha_i$ and terms of the same total degree are ordered with respect to an inverse lexicographic term-ordering.

Example 10 Let $x = (x_1, \ldots, x_s)$ be the indeterminates and let p be a positive integer smaller than s. The p-th elimination ordering is a term-ordering where any monomial containing x_j for a $j \leq p$ is smaller than any monomial not containing any x_j for all $j > p$. An example is $\texttt{lex}$, above. Elimination term-orderings are extensively used in elimination theory.

A term-ordering can be reduced to a $\texttt{lex}$ ordering using the fact that each ordering corresponds to a (non unique) array $M(\tau)$ of integer vectors. The theory of term-ordering classification via matrices is developed in Robbiano (1985), and Adams and Loustaunau (1994). Here we simply state that for the term-orderings of interest in this work the matrix $M(\tau)$ has full rank and for all monomials x^α the condition $x^\alpha \succ_\tau 1$ is equivalent to the fact that in every column of $M(\tau)$ the first non-null entry is positive.

The matrix $M(\tau)$ together with the log function determine τ as follows: for a given term-ordering we have that $\succ_\tau x^\alpha \succ_\tau x^\beta$ if and only if

$$M(\tau) \cdot \alpha \succ_{\texttt{lex}} M(\tau) \cdot \beta$$

referring to the lexicographic ordering over $\mathbb{Z}_+^s$. That is if and only if the first non-null component of $M(\tau) \cdot (\alpha - \beta)$ is positive. Notice that because of

Table 2.1 *Term-orderings in three dimensions. Initial ordering is* $x_1 \succ x_2 \succ x_3$.

Name	Matrix	Polynomial
lex	$\begin{bmatrix} 1 & 0 & 0 \\ 0 & 1 & 0 \\ 0 & 0 & 1 \end{bmatrix}$	$x_1^2 x_2 + x_1 x_2^2 + x_1 x_2 x_3 - x_1^{\dagger}$
deglex	$\begin{bmatrix} 1 & 1 & 1 \\ 1 & 0 & 0 \\ 0 & 1 & 0 \end{bmatrix}$	$x_1^2 x_2 + x_1 x_2^2 + x_1 x_2 x_3 - x_1^{\dagger}$
tdeg	$\begin{bmatrix} 1 & 1 & 1 \\ 0 & 0 & -1 \\ 0 & -1 & 0 \end{bmatrix}$	$x_1^2 x_2 + x_1 x_2^2 + x_1 x_2 x_3 - x_1^{\dagger}$
elimination ordering of x_2	$\begin{bmatrix} 0 & 1 & 0 \\ 1 & 0 & 1 \\ 0 & 0 & -1 \end{bmatrix}$	$x_1 x_2^2 + x_1^2 x_2 + x_1 x_2 x_3 - x_1^{\dagger}$

† The polynomials are written with respect to the ordering in the same row.

the matrix $M(\tau)$ ordering monomials reduces to a collection of inequalities.

Definition 9 *A* **grading** *is a mapping from the monomials* T^s *to non-negative integers* $\mathbb{Z}_+$, *additive in the exponents and such that the grading of 1 is zero.*

Example 11

(i) The total degree, sum of exponents, is a grading defined by $\mathrm{grad}(x^\alpha) = \sum_{i=1}^{s} \alpha_i$.

(ii) The partial degree grading is defined by $\mathrm{grad}(x^\alpha) = \sum_{i \in I} \alpha_i$ where $I \subset \{1, \ldots, s\}$.

If the first row of the matrix associated to a term-ordering τ is the vector $(1, \ldots, 1)$, then τ is a total degree-compatible term-ordering, that is it takes into account the total degree of terms. In general the first row of a term-ordering matrix defines a grading with respect to which the term-ordering is compatible. Table 2.1 shows the most important term-orderings for three indeterminates.

Example 12 [Block orderings] Let $M(\tau)$ be a matrix for the term-ordering τ on $k[x_1, \ldots, x_h]$ and let $N(\sigma)$ be the matrix for the term-ordering σ on

$k[y_1, \ldots, y_i]$. Then, the matrix

$$\begin{bmatrix} M(\tau) & 0 \\ 0 & N(\sigma) \end{bmatrix}$$

represents a term-ordering on $k[x_1, \ldots, x_h, y_1, \ldots, y_i]$ such that $x^\alpha \succ y^\beta$ for all α and β.

Definition 10 *Let τ be a term-ordering on $k[x]$ and f a polynomial in $k[x]$. The leading term of f, $\mathrm{LT}_\tau(f)$ is the largest term with respect to τ among the terms in f. The leading coefficient, $LC_\tau(f)$ is the coefficient of $\mathrm{LT}_\tau(f)$. The leading monomial, $LM_\tau(f)$ is the product $LC_\tau(f)\mathrm{LT}_\tau(f)$.*

Example 13 For $f = 3x_1x_2^2x_3 + \frac{1}{2}x_1x_2x_3^2 + 3x_1^2 \in Q[x_1, x_2, x_3]$ with $\mathtt{tdeg}(x_2 \succ x_1 \succ x_3)$ we have $\mathrm{LT}(f) = x_1x_2^2x_3$ and with $\mathtt{lex}(x_1 \succ x_2 \succ x_3)$ we have $\mathrm{LT}(f) = x_1^2$. In both cases the leading coefficient is 3.

A term-ordering $\succ$ on the terms of $k[x]$ induces naturally a pre-total ordering on the set of all polynomials $k[x]$ given by the leading terms: for $f, g \in k[x]$ we have $f \geq g$ if $\mathrm{LT}(f) \succ \mathrm{LT}(g)$.

2.4 Division algorithm

The operations over $k[x]$ we mostly use are sum, product with scalar, product of polynomials and polynomial division, and in particular simplification of monomial fractions. The first three operations are natural while polynomial division still needs to be discussed and requires the notion of term-ordering. In the univariate case division and the division algorithm for polynomials are well known and summarised in the following theorem.

Theorem 1 *For every pair of polynomials, f and g in one indeterminate, there exist unique polynomials, $s_g, r \in k[x]$ such that $\mathrm{LT}(r) \prec \mathrm{LT}(g)$ and $f = s_g g + r$, where the leading terms are with respect to the only term-ordering in one dimension. The division algorithm returns s_g and r.*

In more than one dimension the situation is more complex.

Theorem 2 *Let $f, g_1, \ldots, g_t$ be in $k[x]$ and τ a term-ordering. There exist $s_1, \ldots, s_t \in k[x]$ and $r \in k[x]$ such that*

$$f = \sum_{i=1}^{t} s_i g_i + r$$

and $\mathrm{LT}_\tau(r)$ is not divisible by any of the $\mathrm{LT}_\tau(g_i)$ for $i = 1, \ldots, t$.

Proof. See Cox, Little and O'Shea (1997, Theorem 3 Chapter 2). See also Section 2.12. □

Definition 11 *The polynomial r of the previous two theorems is called the* remainder. *Sometimes instead of r we write* $\text{Rem}(f, \{g_1, \ldots, g_t\})$ *or* $\text{Rem}(f, G)$ *where G is a finite set of polynomials: $G = \{g_1, \ldots, g_t\}$.*

The sum $\sum_{i=1}^{t} s_i g_i$ is an element of the ideal generated by the g_i's. Neither s_i or r are uniquely defined. Indeed in more than one dimension the division is not a proper operation over the polynomial ring since, in general, its output is not unique, as the following example shows

$$x^2 y + xy^2 + y^2 = (x+1)(y^2 - 1) + x(xy - 1) + 2x + 1$$

giving $r = 2x + 1$ if we divide first by $y^2 - 1$ and

$$x^2 y + xy^2 + y^2 = (x+y)(xy - 1) + (y^2 - 1) + x + y + 1$$

giving $r = x + y + 1$ if we divide first by $xy - 1$.

The division decomposes a polynomial in two parts: the first part has to do with the description of ideals, in particular, design ideals (see Section 3.1), and the second part, the remainder, has to do with the reduction of models.

Over the variety generated by $g_1, \ldots, g_t$, $f = r$. Indeed r is a representation of f over the variety generated by the g_i's but of lower algebraic complexity with respect to the given term-ordering, since the leading term of r is smaller than the leading term of f.

When $r = 0$ then f belongs to the ideal I generated by the g_i's, and this is the solution to the so-called ideal membership problem, standard in algebraic geometry. Unfortunately, as the remainder is not unique, if a computed r is not zero, we cannot conclude whether f belongs to the ideal I. Then, a complete solution would involve the computation of all possible remainders. But we shall see that we can avoid this by introducing a special class of bases for ideals called Gröbner bases for which the remainder is unique regardless of the computation. Even in this case the s_i's remain non unique.

2.5 Hilbert basis theorem

We start with a particular type of ideal which has independent interest.

Definition 12 *A* monomial ideal *is a polynomial ideal I generated by a (possibly infinite) set of monomials, in the multi-index notation*

$$I = \langle x^\alpha : \alpha \in A \subset \mathbb{Z}_+^d \rangle$$

Example 14 The set generated by the leading terms of an ideal is a monomial ideal and we write $\langle \text{LT}(I) \rangle = \langle \text{LT}(f) : f \in I \rangle$.

Example 15 As another example consider the ideal $I = \langle x_1^4 x_2^6, x_1^5 x_2^5, x_1^6 \rangle$ in $k[x_1, x_2]$. An element of I is a k-linear combination of the monomials in the right side of the *leading edge* in Figure 2.1.

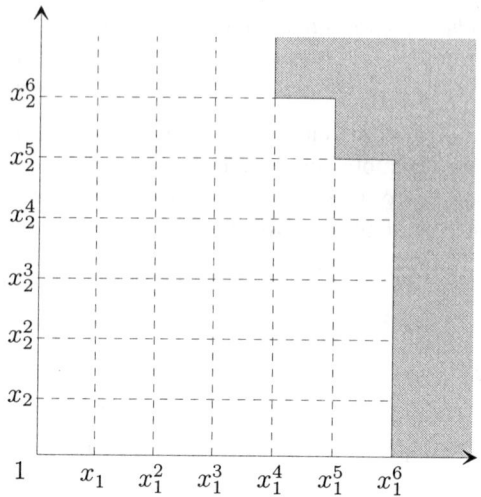

Figure 2.1 *Example of monomial ideal.*

Theorem 3 (Dickson's Lemma) *Every monomial ideal has a finite basis of monomials.*

Proof. The proof of this important theorem can be found in any introductory book of algebraic geometry, see for example Cox, Little and O'Shea (1997). □

A very important consequence of the Dickson's Lemma is its extension to polynomial ideals known as the Hilbert basis theorem. The proof will accustom us to the language and techniques of algebraic geometry.

Theorem 4 (Hilbert Basis Theorem) *Every ideal in $k[x]$ has a finite basis.*

Proof. Let a term-ordering on $k[x]$ be fixed. The set $\langle \mathrm{LT}(I) \rangle = \langle \mathrm{LT}(g) : g \in I \rangle$ is a monomial ideal. The Dickson's lemma states that there exist t polynomials $g_1, \ldots, g_t \in I$ such that the ideal generated by their leading terms is the ideal generated by the leading terms of I, that is $\langle \mathrm{LT}(I) \rangle = \langle \mathrm{LT}(g_1), \ldots, \mathrm{LT}(g_t) \rangle$.

We want to prove that $\langle g_1, \ldots, g_t \rangle = I$. Clearly $\langle g_1, \ldots, g_t \rangle \subseteq I$. We prove the converse by contradiction. From the division algorithm $f \in I$ can be written as $f = \sum_{i=1}^{t} \alpha_i g_i + r$ where r is not divisible by any of $\mathrm{LT}(g_i)$, $i = 1, \ldots, t$. But also $r = f - \sum_{i=1}^{t} \alpha_i g_i \in I$ thus $\mathrm{LT}(r) \in \langle \mathrm{LT}(g_1), \ldots, \mathrm{LT}(g_t) \rangle$, which is a contradiction. Thus $I \subseteq \langle g_1, \ldots, g_t \rangle$. □

We anticipate the fact that the basis found in the proof of Theorem 4 is a Gröbner basis.

2.6 Varieties and equations

Varieties are the geometric counterparts of polynomial ideals. As we have already noticed, a system of polynomial equations is associated with a variety and with an ideal. The link between varieties and ideals is so strict that many problems arising in the context of varieties can be analyzed using techniques from ideal theory and vice versa.

Since any polynomial in the ideal generated by $f_1, \ldots, f_r \in k[x]$ vanishes on the elements of $\{x \in k[x] : f_1(x) = \ldots = f_r(x) = 0\}$, the following definition is well posed.

Definition 13

(i) Let S be a subset of k^s. The set of polynomials defined by

$$\text{Ideal}(S) = \{f \in k[x_1, \ldots, x_s] : f(a_1, \ldots, a_s) = 0 \\ \text{for all } (a_1, \ldots, a_s) \in S\}$$

is an ideal called ideal of S.

(ii) The variety generated by a polynomial ideal $I \subseteq k[x_1, \ldots, x_s]$ is

$$\text{Variety}(I) = \{(a_1, \ldots, a_s) \in k^s : f(a_1, \ldots, a_s) = 0 \text{ for all } f \in I\}$$

A subset of k^s which is a variety of a polynomial ideal in $k[x_1, \ldots, x_s]$ is called a variety.

For $I = \langle f_1, \ldots, f_s \rangle$ we write $\text{Variety}(I)$ or $\text{Variety}(f_1, \ldots, f_s)$. This notation is consistent with the previous one because a point is a zero for the system of polynomial equations if and only if it is a zero for all the polynomials in the ideal generated by the system of polynomial equations.

Another way to describe $\text{Ideal}(S)$ is as the set of all polynomials interpolating the value zero at the points in S. To select one of these polynomials of minimum degree (in some sense) we need a term-ordering. The choice of a term-ordering is a major issue in multi-dimensional interpolation. To understand this point, note that there are many one-dimensional curves through three points in general position in three dimensions. For example both the following curves pass through the points $(1,1,0), (0,0,0), (1,0,1)$

$$\begin{cases} x_1 - x_2 - x_3 = 0 \\ x_2^2 - x_2 = 0 \\ x_3^2 - x_3 = 0 \\ x_2 x_3 = 0 \end{cases} \qquad \begin{cases} x_3 + x_2 - x_1 = 0 \\ x_2^2 - x_2 = 0 \\ x_2 x_1 - x_2 = 0 \\ x_1^2 - x_1 = 0 \end{cases} \tag{2.1}$$

Let S be a set and consider $\text{Ideal}(S)$. In general the variety of such an ideal, $\text{Variety}(\text{Ideal}(S))$, is bigger than S and coincides with S if and only if S is a variety. It is called the *closure* of S, $\bar{S}$ as it is the smallest variety containing S. For example for $S = \{(a_1, a_1) \in \mathbb{R}^2 : a_1 > 0\}$, $\text{Ideal}(S)$ is the ideal generated by $\langle x_2 \rangle \in \mathbb{R}[x_1, x_2]$ and $\text{Variety}(\text{Ideal}(S)) = \{(a_1, a_1) \in \mathbb{R}^2 : a_1 \in \mathbb{R}\}$. This shows that in general subsets of k^s described by inequalities cannot be represented as varieties.

Theorem 5 *The ideal I generated by a set is radical.*

Proof. This is a simple check. □

In some cases Theorem 5 can be used to compute the radical of an ideal as the following univariate example shows.

Example 16 Let us consider the ideal I generated by $f = (x^2 - 4)^3(x^2 + 1)(x^3 - 7)^2$ in $\mathbb{Q}[x]$. The generator f is not square-free and using the unique factorisation of polynomials it is easy to show that the radical ideal of I is generated by $(x^2-4)(x^2+1)(x^3-7)$. On the other side the variety $V(f)$ over the rationals is $\{-2, 2\}$ and the Ideal (Variety(f)) is generated by $x^2 - 4$. Thus the ideal $\langle (x^2 - 4)(x^2 + 1)(x^3 - 7) \rangle$ is strictly bigger than $\sqrt{I}$. Over the algebraically closed field $\mathbb{C}$ all the roots of the equation $f = 0$ belong to the variety generated by f and the equality Ideal (Variety(f)) = $\sqrt{\text{Ideal}(f)}$ holds. We have the following famous theorem.

Theorem 6 (Strong Nullstellensatz) *For a polynomial ideal*

$$\sqrt{I} \subset \text{Ideal}(\text{Variety}(I))$$

When the coefficient field is algebraically closed Ideal (Variety(I)) = $\sqrt{I}$. *In particular* Ideal (Variety(I)) *is radical.*

Proof. See Cox, Little and O'Shea (1997). □

The Strong Nullstellensatz theorem implies that two ideals generate the same variety if and only if their radicals are equal. For the importance of the role played by the hypothesis of algebraic closure see Example 16.

Example 17 [Continuation of Example 16] Any polynomial ideal with complex coefficients $I = \langle f_1, \ldots, f_s \rangle \subset \mathbb{C}[x]$ is generated by one element (that is $\mathbb{C}[x]$ is principal) and in particular $f = \text{GCD}(f_1, \ldots, f_s)$, the greatest common divisor, is a basis for I. Since $\mathbb{C}$ is algebraically closed, f is uniquely factorised as $f = c(x - a_1)^{r_1} \cdots (x - a_p)^{r_p}$ for some $c, a_i \in \mathbb{C}$ and $r_i \in \mathbb{Z}_+$, $i = 1, \ldots, p$. The square-free part of f is $f_{red} = c(x - a_a) \cdots (x - a_p)$. Then Variety$(f_1, \ldots, f_s)$ = Variety(f) = $\{a_1, \ldots, a_s\}$ and the Strong Nullstellensatz theorem becomes

$$\text{Ideal}(\text{Variety}(f_1, \ldots, f_s)) = \text{Ideal}(\text{Variety}(f)) = \langle f_{red} \rangle$$

The above sections allow us to develop further the notion of the model introduced in Definition 4.

Definition 14 *Given a model $f_1, \ldots, f_q, h_1, \ldots, h_l$ the model ideal is the ideal of the model variety and the design ideal is the ideal of the design variety. Alternative terminology is support ideal and ideal of points.*

Definition 15 *Two models are algebraically equivalent if they generate the same variety.*

Above we discussed that the converse to definition 15 is not true in general.
Example 18 [Continuation of Example 4] The model
$$f_1 = bxy + ay - x$$
$$h_1 = x^2 - x$$
generates the variety
$$S = \{(0,0), (1, \frac{1}{a+b})\} \subset \mathbb{Q}(a,b)^2, \quad a \neq 0$$
The ideal generated by S is
$$\langle (a+b)y - x, x^2 - x \rangle \subset \mathbb{Q}(a,b)[x,y]$$
Again f_1 with the constraint $h_1 = x^2 - 1$ gives the variety
$$S = \{(-1, \frac{1}{a-b}), (1, \frac{1}{a+b})\} \subset \mathbb{Q}(a,b)^2$$
In turn, S generates the model
$$\langle (a^2 - b^2)y + bx - a, x^2 - 1 \rangle$$
In both cases the model obtained is explicit.

In the previous discussion we warned that the initial description of the model as in Definition 4 was not suitable for generating the model ideal and the design ideal. It is necessary in some cases to compute a new basis, that is an equivalent model, generating the same variety in order to generate the appropriate ideal. This is done in the following sections.

In our interpretation, the design set of a model is a variety. This set can be either a continuous surface or a finite set of design points. A finite set of points is always a variety. Such a variety is called a zero-dimensional variety. In Chapter 3 we see how to build the ideal for set of points as experimental designs.

We note, without further explanation beyond the scope of the book, that the dimension of a finite set of points is zero, the dimension of a curve is one and of a surface is two. For the exact definition of dimension of a variety we refer to the standard texts, for example Cox, Little and O'Shea (1997).

2.7 Gröbner bases

The Hilbert basis theorem states that any ideal is finitely generated, even if the generating set is not necessarily unique. After the proof of Theorem 4 we mentioned that the basis found was of a special type called Gröbner basis, which we introduce now. The concept of leading term is again essential.

Definition 16 *Let τ be a term-ordering on $k[x]$. A subset $G = \{g_1, \ldots, g_t\}$ of an ideal I is a Gröbner basis of I with respect to τ if and only if*
$$\langle \mathrm{LT}_\tau(g_1), \ldots, \mathrm{LT}_\tau(g_t) \rangle = \langle \mathrm{LT}_\tau(I) \rangle$$

where $\text{LT}_\tau(I) = \{\text{LT}_\tau(f) : f \in I\}$.

In general the following inclusion holds
$$\langle \text{LT}(g_1), \ldots, \text{LT}(g_t) \rangle \subseteq \langle \text{LT}(I) \rangle$$
and unless $\{g_1, \ldots, g_t\}$ is a Gröbner basis, the inclusion may be strict. Indeed in $\langle x_1^3 - 2x_1x_2, x_1^2x_2 - 2x_2^2 + x_1 \rangle \subset \mathbb{Q}[x_1, x_2]$ with the tdeg$(x_1 \succ x_2)$ ordering we have that $x_1^2 \in \langle \text{LT}(I) \rangle$ but $x_1^2 \notin \langle \text{LT}(x_1^3 - 2x_1x_2), \text{LT}(x_1^2x_2 - 2x_2^2 + x_1) \rangle = \langle x_1^3, x_1^2x_2 \rangle$.

Example 19 Clearly a Gröbner basis of the one-dimensional ideal $I = \langle f_1, \ldots, f_t \rangle$ is the greatest common divisor of $f_1, \ldots, f_t$ with respect to the only term-ordering for one-dimensional monomials: for $n, m \in \mathbb{Z}_+$ $x^n \succ x^m$ if and only if $n > m$.

Theorem 7 *Given a term-ordering, every ideal except $\{0\}$ has a Gröbner basis and any Gröbner basis is a basis.*

Proof. This follows from the proof of Theorem 4. □

In particular Gröbner bases are finite sets of polynomials.

In the definition of a Gröbner basis we cannot relax the requirement for a fixed term-ordering.

Example 20 The Gröbner basis of $\langle x_1^2 - 2x_1x_3 + 5, x_1x_2^2 + x_2x_3^3, 3x_2^2 - 8x_3^3 \rangle \in \mathbb{Q}[x_1, x_2, x_3]$ with respect to lex $(x_2 \succ x_1 \succ x_3)$ is

$$\begin{cases} 3x_2^2 - 8x_3^3, \\ 80x_2x_3^3 - 3x_3^8 + 32x_3^7 - 40x_3^5 \\ x_1^2 - 2x_1x_3 + 5 \\ -96x_3^7 + 9x_3^8 + 120x_3^5 + 640x_3^3 x_1 \\ 240x_3^6 + 1600x_3^3 - 96x_3^8 + 9x_3^9 \end{cases}$$

and with respect to tdeg$(x_2 \succ x_1 \succ x_3)$ is

$$\begin{cases} x_1^2 - 2x_1x_3 + 5 \\ -3x_2^2 + 8x_3^3 \\ 8x_1x_2^2 + 3x_2^3 \end{cases}$$

Moreover an ideal can have different Gröbner bases with respect to the same ordering. Both $\{x_2^2 - x_2x_1, x_1^2\}$ and $\{x_2^2 - x_1x_2 + x_1^2, x_1^2\}$ are Gröbner bases with respect to tdeg$(x_2 \succ x_1)$ of the same ideal. We shall see that given a term-ordering an ideal has a unique *reduced* Gröbner basis (see Definition 18).

Definition 17 *A minimal Gröbner basis is a Gröbner basis such that (i) $LC(g) = 1$ for all $g \in G$ and (ii) for all $g \in G$, $\text{LT}(g)$ does not lie in $\langle \text{LT}(G \setminus \{g\}) \rangle$.*

A minimal Gröbner basis is minimal in the sense that none of its proper subsets is a basis for the same ideal.

Definition 18 *A reduced Gröbner basis is a Gröbner basis such that (i) $LC(g) = 1$ for all $g \in G$ and (ii) for all $g \in G$, no term of g lies in $\langle \operatorname{LT}(G \setminus \{g\}) \rangle$.*

Basically, any term of any polynomial in a reduced Gröbner basis is essential. Clearly a reduced Gröbner basis is minimal.

2.8 Properties of a Gröbner basis

The link between Gröbner bases and the division algorithm is expressed by the following theorem. Again, the proof is instructive for the techniques of algebraic geometry we use.

Theorem 8 *Let $I \subset k[x]$ be an ideal, τ a term-ordering, $G = \langle g_1, \ldots, g_t \rangle$ a Gröbner basis for I and $f \in k[x]$. Then, there exist a unique remainder $r \in k[x]$ and a polynomial $g \in I$ such that (i) $f = g + r$ and (ii) no term of r is divisible by one of $\operatorname{LT}(g_1), \ldots, \operatorname{LT}(g_t)$.*

Proof. The existence of g and r follows from the division algorithm with respect to the Gröbner basis (see Section 2.12). The uniqueness is proved by contradiction. Let $f = r_1 + g_1 = r_2 + g_2$, then $r_1 - r_2 = g_2 - g_1 \in I$. In particular $\operatorname{LT}(r_1 - r_2) \in \langle \operatorname{LT}(I) \rangle = \langle \operatorname{LT}(g_1), \ldots, \operatorname{LT}(g_t) \rangle$ since G is a Gröbner basis. That is $\operatorname{LT}(r_1 - r_2)$ is divisible by some of the $\operatorname{LT}(g_i)$ but this is impossible since no term of r_1 and no term of r_2 has this property. Thus $r_1 - r_2 = 0$ and $r_1 = r_2$. The uniqueness of g follows from the uniqueness of r. □

Unfortunately the uniqueness of the remainder r does not imply the uniqueness of the decomposition over a Gröbner basis as the following example shows.

Example 21 *The set $\{x_2 - x_3, x_1 + x_3\}$ is a Gröbner basis in $k[x_1, x_2, x_3]$ with respect to any ordering (Gröbner bases with this property are called* total *Gröbner bases). The following two identities prove the assertion*

$$x_1 x_2 = x_2(x_1 + x_3) + (-x_3)(x_2 - x_3) + (-x_3^2)$$
$$x_1 x_2 = x_1(x_2 - x_3) + (+x_3)(x_1 + x_3) + (-x_3^2)$$

Given a term-ordering, a τ-Gröbner basis and a polynomial, the remainder in Definition 11 is called *normal form*. It is computed in Maple and CoCoA by the command `normalf` and `NF`, respectively.

A most important consequence of Theorem 8 is the ideal membership test.

Corollary 1 *Let I be an ideal in $k[x]$, G be a Gröbner basis of I and f a polynomial in $k[x]$. Then, $f \in I$ if and only if $\operatorname{Rem}(f, G) = 0$.*

Theorem 9 *Given a term-ordering, any non-empty ideal I of $k[x]$ has a unique reduced Gröbner basis.*

Proof. To prove the uniqueness consider two reduced Gröbner bases of I, G_1 and G_2. By definition (in particular point (ii) in Definition 18) they have the same number of elements and the sets of their leading terms coincide: $\text{LT}(G_1) = \text{LT}(G_2)$. In particular for all $g_1 \in G_1$ there exists $g_2 \in G_2$ such that $\text{LT}(g_1) = \text{LT}(g_2)$ and this is a one-to-one correspondence. We have to prove $g_1 = g_2$. On the one hand, since $g_1 - g_2 \in I$ we have $\text{Rem}(g_1 - g_2, G_1) = 0$. On the other hand, none of the terms in $g_1 - g_2$ is divisible by any of the leading terms of the G_1. This implies $\text{Rem}(g_1 - g_2, G_1) = g_1 - g_2$ and concludes the proof of the uniqueness of reduced Gröbner bases.

For the proof of the existence of reduced Gröbner bases see Cox, Little and O'Shea (1997). □

The following theorem lists features of Gröbner bases we shall refer to.

Theorem 10 *Let I be an ideal in $k[x]$ and τ a term-ordering. The following statements are equivalent.*

1. $G = \{g_1, \ldots, g_t\}$ *is a Gröbner basis for I.*

2. *For all $f \in I \setminus \{0\}$ there exists an element $g_i \in G$ such that $\text{LT}(g_i)$ divides $\text{LT}(f)$.*

3. $\langle \text{LT}(I) \rangle = \langle \text{LT}(G) \rangle$.

4. *For all $f \in I$ we have $\text{Rem}(f, G) = 0$ (ideal membership).*

5. *Any element $f \in I$ is decomposed over G in the following way*

$$f = \sum_{i=1}^{t} f_i g_i \quad \text{and} \quad \text{LT}(f) = \max(\text{LT}(f_i)\,\text{LT}(g_i))$$

6. *For any polynomial $f \in k[x]$ there exists unique $\text{Rem}(f, G)$ (remainder theorem).*

Proof. For the proof see Adams and Loustaunau (1994). □

We conclude this section with the important remark that Gröbner bases are computational objects. The first algorithm to compute Gröbner bases is due to Buchberger (1966). Independently in 1964 H. Hironaka described a similar concept to Gröbner bases he called "standard bases." In Section 2.12.3 we present a version of the Buchberger algorithm.

2.9 Elimination theory

Gröbner bases with the `lex` ordering are especially used to solve systems of polynomial equations. The more general use of Gröbner bases to solve systems of equations is called elimination theory. A main theorem used in elimination theory is the Weak Nullstellensatz.

Theorem 11 (Weak Nullstellensatz) *Let k, the coefficient field, be algebraically closed, then* Variety$(I) = \emptyset$ *if and only if* $I = k[x]$.

Proof. See Cox, Little and O'Shea (1997). □

The problem of whether a system of polynomial equations $f_1 = \ldots = f_v = 0$ has a solution is called the *consistency* problem. In terms of varieties this is equivalent to asking whether the variety Variety$(f_1, \ldots, f_v)$ is empty. Because of the Weak Nullstellensatz Theorem and the uniqueness of reduced Gröbner bases, over algebraically closed fields there is no solution to the system if and only if the reduced Gröbner basis (with respect to any term-ordering) of the polynomial ideal associated to the system is $\{1\}$. In a non-algebraically closed field the condition is sufficient but not necessary, a counter example is that $1 + x^2 = 0$ has no solution in $\mathbb{R}$ but $\{x^2 + 1\}$ is a reduced Gröbner basis.

Theorem 12 *Let τ be a p-th elimination term-ordering of the indeterminates $x_{p+1}, \ldots, x_s$. Let $g_1, \ldots, g_t$ be a Gröbner basis of the ideal I. Then*

$$\{g_1, \ldots, g_t\} \bigcap k[x_{p+1}, \ldots, x_s]$$

is a Gröbner basis of the elimination ideal $I \bigcap k[x_{p+1}, \ldots, x_s]$.

Proof. See Cox, Little and O'Shea (1997). □

The elimination theory implies that to solve the system of polynomial equations $f_1 = \ldots = f_v = 0$ one can first find a reduced Gröbner basis of the ideal $\langle f_1, \ldots, f_v \rangle$ with respect to the `lex` ordering. Call this $\{g_1, \ldots, g_t\}$. The new system of equations $g_1 = \ldots = g_t = 0$ has essentially a triangular form, which can be solved by backward substitution, and the two systems have the same solutions since they generate the same ideal.

Example 22 With respect to the term-ordering `lex`$(x_1 \succ x_2 \succ x_3)$ the Gröbner basis of $f_1 = x_1^2 - x_2^2, f_2 = x_1 + x_2^2 + x_3^2, f_3 = x_3 + x_1 + x_2$ is

$$\begin{cases} g_1 = x_1 + x_2 + x_3 \\ g_2 = x_2^2 + x_3^2 - x_2 - x_3 \\ g_3 = x_3^2 + 2x_2 x_3 \\ g_4 = -2x_3^2 + 5x_3^3 \end{cases}$$

And the system $f_1 = f_2 = f_3 = 0$ can be solved by solving the triangular system $g_1 = g_2 = g_3 = g_4 = 0$.

Other examples of applications of elimination theory will be given in Chapter 3 where it is an essential tool in the construction of design ideals. Let us now consider the variety associated to the projection ideal.

Theorem 13 *Let V be a variety and* $\text{Ideal}(V)$ *its ideal. Let π be the projection of k^s onto the last coordinates $x_{p+1}, \ldots, x_s$. Then*

$$\text{Ideal}(\pi(V)) = \text{Ideal}(V) \bigcap k[x_{p+1}, \ldots, x_s]$$

Proof. Let $g_1, \ldots, g_t$ be a Gröbner basis of $\text{Ideal}(V)$ with respect to a term-ordering as in Theorem 12. Then, the basis has the structure

$$g_1(x_{p+1}, \ldots, x_s)$$
$$\vdots$$
$$g_l(x_{p+1}, \ldots, x_s)$$
$$g_{l+1}(x_1, \ldots, x_s)$$
$$\vdots$$
$$g_t(x_1, \ldots, x_s)$$

where $g_1, \ldots, g_l$ is a Gröbner basis of $\text{Ideal}(V) \bigcap k[x_{p+1}, \ldots, x_s]$ and the polynomials $g_1, \ldots, g_t$ form a Gröbner basis of $\text{Ideal}(V)$. If $a \in \pi(V)$ and $a = (a_1, \ldots, a_s) \in V$ then $g_i(a_{p+1}, \ldots, a_s) = 0$ for $i = 1, \ldots, l$ and all elements of $\text{Ideal}(V) \bigcap k[x_{p+1}, \ldots, x_s]$ are zero on such points. This shows that

$$\text{Ideal}(\pi(V)) \subset \text{Ideal}(V) \bigcap k[x_{p+1}, \ldots, x_s]$$

Now assume $f \in k[x_{p+1}, \ldots, x_s]$ is in $\text{Ideal}(V)$. Then, for all $a \in V$, $f(a) = 0$ so that $f \in \text{Ideal}(V)$. Note that $\pi(V) = \text{Variety}(\text{Ideal}(\pi(V))) = \text{Variety}(\text{Ideal}(V) \cap k[x_{p+1}, \ldots, x_s])$. □

Example 23 The reduced Gröbner basis for the model in Example 4 and with respect to the term-ordering $\texttt{lex}\,(t \succ x \succ y)$ is

$$\begin{cases} (yb-1)(x-c) + ya \\ t - x + c \end{cases}$$

where the polynomial relationship linking input (x) and output (y) is given by the first polynomial. The Gröbner basis with respect to the term-ordering $\texttt{lex}\,(x \succ y \succ t)$ giving the representation of the model remains the same.

Example 24 For Example 5 the first two polynomials in the Gröbner basis with respect to lex $(t_1 \succ t_2 \succ y \succ x_1 \succ x_2)$ give the input/output relation

$$\begin{cases} x_1{}^2 + x_2{}^2 - 1, \\ 2y^2b^2 + 2a^2b^2 - 2a^2r^2 + y^4 + r^4 - 2b^2r^2 - 2y^2r^2 + 4x_2{}^2y^2a^2 + a^4 \\ \quad - 2y^2a^2 + b^4 - 8aby^2x_1x_2 - 4x_2{}^2y^2b^2, \\ ax_1x_2y^2 + ax_1x_2b^2 + x_1x_2a^3 - ax_1x_2r^2 + 2a^2t_2x_2{}^2 - 2a^2t_2 \\ \quad + 2b^2x_2{}^2t_2 - bx_2{}^2y^2 - b^3x_2{}^2 + bx_2{}^2r^2 - bx_2{}^2a^2, \\ 2at_2x_1 + 2bx_2t_2 - x_2y^2 - x_2b^2 + x_2r^2 - x_2a^2, \\ t_2y^2 + t_2b^2 + a^2t_2 - t_2r^2 - 2ax_1x_2y^2 - 2bx_2{}^2y^2, \\ t_2{}^2 - x_2{}^2y^2, \\ -y^2 + 2t_1a - a^2 + 2t_2b - b^2 + r^2 \end{cases}$$

2.10 Polynomial functions and quotients by ideals

Quotients by ideals play a key role in the algebraic theory of identifiability of Chapter 3. The ring-isomorphism between polynomial functions over a variety and quotients by the variety ideal justifies the theory.

Definition 19 *Let $V \subset k^s$ be a variety. A function*

$$\Phi : V \longrightarrow k$$

is a polynomial function (or mapping) if there exists a polynomial $f \in k[x_1, \ldots, x_s]$ such that

$$\Phi(a_1, \ldots, a_s) = f(a_1, \ldots, a_s)$$

for all points $(a_1, \ldots, a_s)$ in V. The polynomial f is said to represent Φ. The collection of polynomial functions over V is denoted by $k[V]$. The polynomials that represent the same Φ are said to be confounded over V.

Notice that two polynomials f and $g \in k[x]$ represent the same polynomial function on V if and only if $f - g \in \text{Ideal}(V)$.

The set $k[V]$ is an Abelian ring with the following operations:

$$(\Phi + \Psi)(a) = \Phi(a) + \Psi(a)$$
$$(\Phi \cdot \Psi)(a) = \Phi(a) \cdot \Psi(a)$$
$$(\alpha \Phi)(a) = \alpha(\Phi(a))$$

where $\alpha \in k$, $\Phi, \Psi \in k[V]$ and $a \in V$. Moreover if f represents Φ and g represents Ψ then $f + g$ represents $\Phi + \Psi$ and $f \cdot g$ represents $\Phi \cdot \Psi$.

Definition 20 *Let I be an ideal in $k[x]$. The quotient of $k[x]$ modulo I is defined as*

$$k[x]/I = \{[f] : f \in k[x]\}$$

where we define $[f] = \{g \in k[x] \text{ such that } f - g \in I\}$.

The set $k[x]/I$ has the algebraic structure of a k-algebra. For all f and g in I and a scalar α in k we have the following definitions

$$[f] + [g] = [f + g]$$
$$[f][g] = [fg]$$
$$\alpha[f] = [\alpha f].$$

If the ideal I is generated by a variety V, then we have that $f \equiv g$ modulo $\mathrm{Ideal}\,(V)$, that is $[f] = [g]$ if and only if f and g define the same polynomial function V. This connection is exploited in the next theorem.

Theorem 14 *The sets $k[x]/\mathrm{Ideal}\,(V)$ and $k[V]$ are ring isomorphic.*

Proof. See Cox, Little and O'Shea (1997). □

The division algorithm allows us to produce simple representations of equivalence classes for congruence modulo an ideal. Given a Gröbner basis of the ideal, two polynomials f and g are equivalent if and only if they have the same remainder with respect to the Gröbner basis. This congruence is carried over to the set of polynomial functions over the variety defined by the ideal. The next theorem reinterprets the division and the form of the remainder in this context.

Theorem 15 *Let a term-ordering on $k[x]$, τ be fixed and let I be an ideal in $k[x]$.*

1. *Every $f \in k[x]$ is congruent modulo I to a unique polynomial r which is a k-linear combination of the monomials in the complement of the monomial ideal $\langle \mathrm{LT}(I) \rangle$.*

2. *The elements of the set $A = \{x^\alpha : x^\alpha \notin \langle \mathrm{LT}_\tau(I) \rangle\}$ are linearly independent modulo I, that is modulo $\mathrm{Ideal}\,(V)$*

$$\sum_{x \in A} c_\alpha x^\alpha = 0 \qquad \text{with } c_\alpha \in k$$

if and only if $c_\alpha = 0$ for all α.

Proof. See Cox, Little and O'Shea (1997). □

That is $k[x]/I$ is isomorphic as k-vector space to

$$\mathrm{Span}(x^\alpha : x^\alpha \notin \langle \mathrm{LT}(I) \rangle)$$

Notice that the property $x^\alpha \notin \langle \mathrm{LT}(I) \rangle$ is derived finitely by checking that x^α does not divide any of the leading terms of the reduced Gröbner basis of the ideal with respect to τ.

Different term-orderings give different bases for the above set. But they have all the same cardinality since $\mathrm{Span}(x^\alpha : x^\alpha \notin \langle \mathrm{LT}_\tau(I) \rangle)$ are all k-vector spaces isomorphic to $k[x]/I$. For example the left-hand system of polynomial equations in (2.1) gives a Gröbner basis with respect to the lex $(x_1 \succ x_2 \succ x_3)$ term-ordering and the right-hand system is a Gröbner basis with respect to the lex $(x_3 \succ x_2 \succ x_1)$ term-ordering. Of course they represent the same ideal I. In both cases the dimension as a $\mathbb{Q}$-vector space of $\mathbb{Q}[x_1, x_2, x_3]/I$ is 3. On the left-hand side, $\mathbb{Q}[x]/I$ is represented with $\{1, x_3, x_2\}$ and on the right-hand side, with $\{1, x_1, x_2\}$.

The above holds also for nonzero-dimensional ideals. The vector space basis of the quotient ring plays a crucial role in this book. The last result in this chapter characterises zero-dimensional varieties via the structure of ideal bases and the structure of the quotient space.

Theorem 16 *Let τ be a term-ordering and let k be algebraically closed. Let $V = \text{Variety}(I)$ be a variety over $k[x]$ and G a Gröbner basis for I. The following statements are equivalent.*

1. V is finite.

2. For each $i = 1, \ldots, s$ there is $m_i > 0$ and $g \in G$ such that $x_i^{m_i} = LT(g)$.

3. The k-vector space $k[x_1, \ldots, x_s]/I$ is finite-dimensional.

Proof. See Cox, Little and O'Shea (1997). □

By the two previous theorems it follows that

Theorem 17 *The vector space basis $\{x^\alpha : x^\alpha \notin \langle LT(I) \rangle\}$ has N elements if and only if V has N elements.*

Theorem 16 characterises zero-dimensional ideals. In particular it follows that algorithmically the operations over the quotient space $k[x]/I$ can be performed via Gröbner bases and the remainder theorem (Theorem 10, Item 6). For example let G be a Gröbner basis for I then $[f+g] = \text{Rem}(f + g, G)$ modulo I.

2.11 Hilbert function

In this book we will not discuss the notion of dimension of a variety, briefly mentioned in Section 2.6. Nevertheless in Section 3.11 we use the Hilbert function which is at the basis of the definition of dimension of a variety.

Definition 21 *Let I be a polynomial ideal and s a non-negative integer. The set of polynomials in I of total degree less that or equal to s is indicated with $I_{\leq s}$.*

Example 25 For the trivial ideal $I = k[x_1, \ldots, x_d]$, $I_{\leq 1}$ is the set of linear forms in $x_1, \ldots, x_d$ indeterminates.

Definition 22 *Let I be an ideal in $k[x_1, \ldots, x_d]$ and s a non-negative integer. The affine Hilbert function of I is the integer function*

$$^a HF_I(s) = dim(k[x_1, \ldots, x_d]_{\leq s}) - dim(I_{\leq s})$$

where $dim(V)$ is the dimension of the vector space V.

Definition 23 *An ordering $\succ$ such that $\sum_{i=1}^{d} \alpha_i > \sum_{i=1}^{d} \beta_i$ implies $x^\alpha \succ x^\beta$ is called a* graded *ordering.*

Theorem 18 *Let I be an ideal in $k[x_1, \ldots, x_d]$ different from $\{0\}$ and $k[x_1, \ldots, x_d]$. The following holds*

(i) *Let I be a monomial ideal. Then ${}^aHF_I(s)$ is the number of monomials of total degree less than or equal to s in the quotient space $k[x_1, \ldots, x_d]/I$.*

(ii) *For all sufficiently large s the affine Hilbert function of I is the polynomial*

$$^aHF_I(s) = \sum_{i=0}^{d} b_i \binom{s}{d-i}$$

where b_i are integers and b_0 is positive. Such a polynomial is called the affine Hilbert polynomial.

(iii) *If I is a monomial ideal then the degree of the polynomial in (ii) above is the maximum of the dimensions of the coordinate subspaces contained in Variety(I).*

(iv) *For a graded ordering the affine Hilbert function of $\langle \text{LT}(I) \rangle$ and I coincide.*

(v) *The affine Hilbert polynomials of I and its radical have the same degree.*

(vi) *Given a design D with l_i levels in the x_i, the Hilbert function of the design ideal, Ideal(D) is zero for $s > \sum_{i=1}^{d}(l_i - 1)$.*

Proof. For Items (i), (ii), (iii), (iv) and (v) see Cox, Little and O'Shea (1997). Item (vi) follows directly form (i). □

2.12 Further topics

2.12.1 Division algorithm

At each step of the division algorithm in one dimension, the term of maximum degree (the leading term) of the dividend is well defined and it is divided by the leading term of the divisor. The notion of term-ordering makes possible the division algorithm for multi-dimensional polynomials by allowing us to select leading terms. The simplification of monomials is the foundation of the division algorithm. Thus we recall that x^α divides x^β if and only if all the components of $\alpha - \beta$ are greater or equal to 0. See Definition 8, Item 2.

An example shows how the division algorithm works. In $\mathbb{Q}[x_1, x_2, x_3]$ with the $\texttt{tdeg}(x_1 \succ x_2 \succ x_3)$ ordering we want to divide $x_1^3 x_3 x_2^2 + x_2$ by

FURTHER TOPICS 37

Table 2.2 *Division algorithm.*

Input $g_1, \ldots, g_t$ and f
Output $s_1, \ldots, s_t$ and r such that 1. $f = \sum_{i=1}^{t} s_i g_i + r$
                                           2. $\mathrm{LT}(r)$ is not divisible by $\mathrm{LT}(g_i)$
begin $s_1 = s_2 \ldots = s_t := 0$
         $r := 0$
         $p := f$
   while $p \neq 0$ **do**
         $i := 1$
         Division_Occured   := FALSE
         while $i \leq t$ **and**   Divison_Occurred = FALSE **do**
                           if $\mathrm{LT}(g_i)$ divides $\mathrm{LT}(p)$ **then**
                                          $s_i := s_i + \mathrm{LT}(p)/\mathrm{LT}(g_i)$
                                          $p := p - \mathrm{LT}(p)/\mathrm{LT}(g_i) g_i$
                                          Division_Occured := TRUE
                           else
                                          $i := i + 1$
         if Division_Occurred = FALSE **then**
                                          $r := r + \mathrm{LT}(p)$
                                          $p := p - \mathrm{LT}(p)$
end

$x_1 x_2 + x_3, x_1 x_3$ and x_3 in the given sequence. The scheme is as follows

$$s_1 : x_1^2 x_3 x_2 - x_1 x_3^2$$
$$s_2 : x_3^2$$
$$s_3 :$$

$g_1 : x_1 x_2 + x_3$	$x_1^3 x_3 x_2^2 + x_2$
$g_2 : x_1 x_3$	$x_1^3 x_3 x_2^2 + x_3^2 x_1^2 x_2$
$g_3 : x_3$	$- x_1^2 x_3^2 x_2 + x_2$
	$-x_1^2 x_3^2 x_2 - x_1 x_3^3$
	$- x_1 x_3^3 + x_2$
	$-x_1 x_3^3$
	x_2

First we notice that $\mathrm{LT}(x_1^3 x_3 x_2^2 + x_2)$ is divided by $\mathrm{LT}(x_1 x_2 + x_3)$ giving $x_1^2 x_3 x_2$. We multiply it by $x_1 x_2 + x_3$ and subtract from $x_1^3 x_3 x_2^2 + x_2$ getting $x_1^3 x_3 x_2^2 + x_2 = (x_1 x_2 + x_3) x_1^2 x_3 x_2 + (-x_1^2 x_2 x_3^2 + x_2)$. Again $\mathrm{LT}(-x_1^2 x_2 x_3^2 + x_2)$ is divided by $\mathrm{LT}(x_1 x_2 + x_3)$ and we have $x_1^3 x_3 x_2^2 + x_2 = (x_1 x_2 + x_3)(x_1^2 x_3 x_2 - x_1 x_3^2) + (x_1 x_3^3 + x_2)$. Now $\mathrm{LT}(x_1 x_3^3 + x_2)$ is not divisible by $\mathrm{LT}(x_1 x_2 + x_3)$ but by $\mathrm{LT}(x_1 x_3)$ and we have $x_1^3 x_3 x_2^2 + x_2 = (x_1 x_2 + x_3)(x_1^2 x_3 x_2 - x_1 x_3^2) + (x_1 x_3)(x_3^2) + (1)(x_2)$. Since $\mathrm{LT}(x_2)$ is not divisible by any of $\mathrm{LT}(x_1 x_2 + x_3)$, $\mathrm{LT}(x_1 x_3)$ and $\mathrm{LT}(x_3)$, x_2 is the remainder of the division.

Remember that the division is not a properly defined operation since its

Table 2.3 *The algebra-geometry dictionary.*

Geometry	Algebra
V	$\text{Ideal}(V)$
$V \cap W$	$\sqrt{\text{Ideal}(V) + \text{Ideal}(W)}$
$V \cup W$	$\sqrt{\text{Ideal}(V) \cdot \text{Ideal}(W)}$
$V \cup W$	$\text{Ideal}(V) \cap \text{Ideal}(W)$
$\pi_p(\text{Variety}(I))^\dagger$	$\sqrt{I \cap k[x_{p+1}, \ldots, x_s]}$

† π_p is the projector over the last $s - p - 1$ variables.

result, and thus the output of the above algorithm, depends on the order on which the dividends are considered. Gröbner basis theory addresses this issue. The division algorithm is discussed in considerably in the literature and in Table 2.2 we report the code from Cox, Little and O'Shea (1997).

2.12.2 Algebra-geometry dictionary

The connection between varieties and radical ideals is so strong that an algebra-geometry dictionary has been made. The parts of this dictionary we shall use are shown in Table 2.3. They are taken from Cox, Little and O'Shea (1997). We suppose that the ideals involved are radical and the coefficient field is algebraically closed.

2.12.3 The Buchberger algorithm

Before describing the Buchberger algorithm we define S-polynomials (S-poly). In particular S-polynomials are used to test whether a set of polynomials is a Gröbner basis.

Definition 24 *Let f and g be polynomials in $k[x_1, \ldots, x_s]$, τ a term-ordering and $x_1^{\gamma_1} \cdots x_s^{\gamma_s}$ the least common multiple (LCM) of $\text{LT}(f)$ and $\text{LT}(g)$. Then the S-polynomial of f and g is defined as*

$$\text{S-poly}(f, g) = \frac{x_1^{\gamma_1} \cdots x_s^{\gamma_s}}{LM(f)} f - \frac{x_1^{\gamma_1} \cdots x_s^{\gamma_s}}{LM(g)} g$$

Simply, S-poly(f, g) is the mechanism by which we cancel leading terms to produce the decomposition of Item 5 in Theorem 10. It is proved, see Cox, Little and O'Shea (1997, Chapter 2 Lemma 5), that every cancellation of leading terms among polynomials of the same total degree involves S-polynomials. When considering two polynomials at a time, this can be interpreted in a different way to compute the S-polynomial itself.

For example, $x_3(2x_1^3 + x_3) - 2x_1(x_1^2 x_3 + x_2^2) = x_3^2 - 2x_1 x_2^2$ is equal to
$2\,\text{S-poly}(2x_1^3 + x_3, x_1^2 x_3 + x_2 x_3) = \frac{x_1^3 x_3}{2x_1^3}(2x_1^3 + x_3) - \frac{x_1^3 x_3}{x_1^2 x_3}(x_1^2 x_3 + x_2^2)$.

Let us detail how S-polynomials arise in the division algorithm. We want to divide f by $f_1, \ldots, f_r$. In the division algorithm it may happen that both $\text{LT}(f_i)$ and $\text{LT}(f_j)$ divide the leading term X of f for some $i \neq j$. If we divide X by f_i then we have $h_1 = f - \frac{X}{\text{LT}(f_i)} f_i$. If we divide X by f_j then we have $h_2 = f - \frac{X}{\text{LT}(f_j)} f_j$. An ambiguity is introduced, that is the reason why the decomposition of Item 5 in Theorem 10 may not be unique, namely

$$h_2 - h_1 = \frac{X}{\text{LT}(f_i)} f_i - \frac{X}{\text{LT}(f_j)} f_j = \frac{X}{\text{LCM}(\text{LT}(f_j), \text{LT}(f_i))}\,\text{S-poly}(f_i, f_j).$$

Theorem 19 *A basis $G = \{g_1, \ldots, g_r\}$ of an ideal I is a Gröbner basis if and only if for each pair (i, j), $i, j \in \{1, \ldots, r\}$*

$$\text{Rem}(\text{S-poly}(g_i, g_j), G) = 0$$

Proof. See Cox, Little and O'Shea (1997). □

Theorem 19 gives a finite test to verify whether a set of polynomials is a Gröbner basis.

Up to now we have seen that any ideal except $\{0\}$ has a Gröbner basis and that it has a unique reduced Gröbner basis. Table 2.4 follows Chapter 2.7 of Cox, Little and O'Shea (1997) and shows a three-part version of the Buchberger algorithm to compute the reduced Gröbner basis of an ideal given a finite generating set and a term-ordering τ. The first part returns a Gröbner basis for the ideal, the second one makes it minimal and the third one makes it reduced. Consistency, finiteness and correctness of the algorithm are proved in the literature.

Computer algebra packages use versions of the Buchberger algorithm improved by sophisticated programming and additional mathematical ideas (for example, the Gebauer-Möller formulae, see Caboara, de Dominicis and Robbiano (1996), Marinari, Möller and Mora (1996)).

Broadly speaking, the Buchberger algorithm is a generalisation of the Gaussian elimination or row reduction algorithm for linear systems, as the following example shows.

Example 26 The Gauss Jordan elimination of the matrix A below

$$A = \begin{bmatrix} 3 & -6 & -2 & 0 \\ 2 & -4 & 0 & 4 \\ 1 & -2 & -1 & -1 \end{bmatrix}$$

is

$$B = \begin{bmatrix} 1 & -2 & 0 & 2 \\ 0 & 0 & 1 & 3 \\ 0 & 0 & 0 & 0 \end{bmatrix}$$

Table 2.4 *Buchberger algorithm.*

Input $F = (f_1, \ldots, f_r)$
Output $G = (g_1, \ldots, g_t) \supset F$ Gröbner basis
begin
$G := F$
repeat
    $G1 := G$
    for each pair $(p, q) \in G1$ and $p \neq q$ **do**
        $S := \text{Rem}(\text{S-poly}(p, q), G1)$
        if $S \neq 0$ **then** $G := G \cup \{S\}$
until $G = G1$
end

Input $G = (g_1, \ldots, g_t)$ Gröbner basis
Output $M = (m_1, \ldots, m_u)$ minimal Gröbner basis
begin
$M := G$
for all $f \in M$ **do**
    if $\text{LT}(f) \in \langle \text{LT}(M - \{f\}) \rangle$ **then** $M := M - \{f\}$
return M
end

Input $M = (m_1, \ldots, m_u)$ minimal Gröbner basis
Output $R = (r_1, \ldots, r_v)$ reduced Gröbner basis
begin
$R := M$
for all $g \in R$ **do**
    $g1 := \text{Rem}(g, R - \{g\})$
    $R := (R - \{g\}) \cup \{g1\}$
return R
end

The set of polynomials obtained by right multiplication of B with the indeterminate vector (x_1, x_2, x_3, x_4) is

$$G = \{x_1 - 2x_2 + 2x_4, 3x_4 + x_3\}$$

Now consider the set of polynomials, F obtained by right multiplication of A with (x_1, x_2, x_3, x_4)

$$F = \{3x_1 - 6x_2 - 2x_3, 2x_1 - 4x_2 + 4x_4, x_1 - 2x_2 - x_3 - x_4\}$$

The Gröbner basis of F with respect to $\text{lex}(x_1 \succ x_2 \succ x_3 \succ x_4)$ is equal to G above. That is F and G above generate the same ideal since the rows of B are obtained by those of A with ideal operations, and using the S-polynomial test we see that a reduced echelon matrix leads to only one reduced Gröbner basis. For a discussion on the links between Gröbner bases and systems of linear equations we refer to Becker and Weispfenning (1993) and Mora (1994).

In general the computation of the Gröbner basis is very expensive, the cost depends heavily on the term-ordering, the worst being usually the `lex` ordering. There are methods to speed the computation based on the Hilbert function for polynomials (see Cox, Little and O'Shea (1997, Chapter 9)) implemented in the packages we use. For a reference see Traverso (1996).

CHAPTER 3

Gröbner bases in experimental design

In this chapter we use the methods of algebraic geometry to study the identifiability problem in experimental design: given a design which model(s) can we identify? As mentioned the starting point is to represent a design D as a variety, namely the solution of a set of polynomial equations, or equivalently the design ideal, that is, the set of all polynomials interpolating the design points at zero. The principal result is that starting with a class of models M (usually M will be the set of all polynomials in d indeterminates) the quotient vector space $M/\operatorname{Ideal}(D)$ yields a class of identifiable terms. The theory of Gröbner bases is used to characterise the design ideal and the quotient space.

The following problems will be addressed in particular.

(i) Which classes of polynomial models does a given design identify?

(ii) Is a given model identifiable by a given design?

(iii) What is confounding/aliasing in this context?

(iv) What conditions must M satisfy so that the theory applies?

This algebraic approach to identifiability in experimental design was introduced by Pistone and Wynn (1996).

3.1 Designs and design ideals

A design is a zero-dimensional variety, that is a pointwise finite subset without replications, equivalently a single replicate design. Our philosophy is to use in an interchangeable way the representation of a design as a variety and as an ideal.

Definition 25

(i) A design *is a finite set of distinct points in* k^d.

(ii) We call the design ideal *of D (in symbol* $\operatorname{Ideal}(D)$*) the ideal generated by D, that is the set of all polynomials in $k[x_1, \ldots, x_d]$ whose zeros include the design points. Alternative terminology is* support ideal *and* ideal of points.

The following theorem applies the general theory of Chapter 2 to the specific case of designs.

Theorem 20 *Given a design D,*

(i) the design ideal, $\mathrm{Ideal}(D)$ is the intersection of the design ideals generated by the single design points of D.

(ii) The design D is a variety.

(iii) The design ideal $\mathrm{Ideal}(D)$ is a radical ideal.

Proof. The proof of Item (i) gives a method to construct a basis for $\mathrm{Ideal}(D)$ as we shall see later. A design is the union of its single design points. The single point $(a_1, \ldots, a_d)$ is clearly the variety defined by the ideal $\langle (x_1 - a_1), \ldots, (x_d - a_d) \rangle$. Finite unions of varieties correspond to finite intersections of ideals (see Table 2.3). That is $\mathrm{Ideal}\,(V_1 \cup V_2) = \mathrm{Ideal}\,(V_1) \cap \mathrm{Ideal}\,(V_2)$. A basis for the design ideal is given by the products of all elements in bases of the single design points. The intersection of the two design ideals generated by $f_1 = (x_1 - a_1), \ldots, f_d = (x_d - a_d)$ and $g_1 = (x_1 - b_1), \ldots, g_d = (x_d - b_d)$ respectively is $f_i g_j$ for $i, j = 1, \ldots, d$. In a finite number of steps by adjoining a point at the time we obtain a basis B for the design ideal.

One can see that the design D is defined by the solutions of the basis B obtained above and thus D is a variety.

If $f^m(x) = 0$ for all $x \in D$ and some positive m then clearly $f(x) = 0$ for all $x \in D$. This proves that the $\mathrm{Ideal}(D)$ is radical. □

Example 27 The 2^2 full factorial design $\{(\pm 1, \pm 1)\} \subset \mathbb{Q}^2$ corresponds to the ideal of $\mathbb{Q}[x_1, x_2]$ generated by $(x_1 - 1)(x_1 + 1)$ and $(x_2 - 1)(x_2 + 1)$. Indeed the solutions of the system of equations

$$\begin{cases} x_1^2 - 1 = 0 \\ x_2^2 - 1 = 0 \end{cases}$$

are $\{(\pm 1, \pm 1)\}$.

In general a zero-dimensional variety is generated by the radical ideal containing a polynomial interpolating the points. For example, the above design ideal could be represented by the polynomial

$$(x_1^2 - 1)^2 + (x_2^2 - 1)^2$$

This feature will be exploited in Chapter 4.

3.2 Computing the Gröbner basis of a design

Consider the problem of finding the ideal associated to the set of points D where

$$D = \{(a(1)_1, \ldots, a(1)_d), \ldots, (a(N)_1, \ldots, a(N)_d)\}$$

One method which does not give rise to a Gröbner basis is used in the proof of Theorem 20. The algorithm is as follows.

(i) Consider the reduced Gröbner basis of the single point design ideal, $a(i)$, $\langle x_1 - a(i)_1, \ldots, x_d - a(i)_d \rangle$.

(ii) Intersect the single point design ideals of all the design points.

Some computer algebra packages provide built-in procedures to compute Gröbner bases of intersections of ideals. Other methods to find the design ideal return directly its reduced Gröbner basis and thus assume a term-ordering. Next we present a method which is a direct application of elimination theory. We start with an example.

Example 28 [Continuation of Example 27] The considered design corresponds to the projection over $\mathbb{Q}[x_1, x_2]$ of the 6-dimensional ideal in $\mathbb{Q}[t_1, t_2, t_3, t_4, x_1, x_2]$ generated by the following polynomials

$$t_1(x_1 - 1), t_1(x_2 - 1), t_2(x_1 + 1), t_2(x_2 - 1),$$
$$t_3(x_1 - 1), t_3(x_2 + 1), t_4(x_1 + 1), t_4(x_2 + 1),$$
$$t_1 + t_2 + t_3 + t_4 - 1$$

The last polynomial excludes unwanted points given by $t_i = 0$, $i = 1, \ldots, 4$. Next compute the Gröbner basis of the previous nine polynomials with respect to an elimination term-ordering, such as lex, and the initial ordering $t_1 \succ t_2 \succ t_3 \succ t_4 \succ x_1 \succ x_2$. It is

$$G = \{4t_1 - x_2 x_1 - x_1 - 1 - x_2,$$
$$4t_2 + x_2 x_1 - x_1 + x_2 - 1,$$
$$4t_3 + x_1 - 1 + x_2 x_1 - x_2,$$
$$4t_4 - x_2 x_1 + x_2 + x_1 - 1,$$
$$x_1^2 - 1, x_2^2 - 1\}$$

The elements of G not containing the t_i's form the sought Gröbner basis of $\text{Ideal}(D)$, namely

$$\{x_1^2 - 1, x_2^2 - 1\}$$

The generalisation to d-dimensions is as follows. The N-point variety

$$\{(a(1)_1, \ldots, a(1)_d), \ldots, (a(N)_1, \ldots, a(N)_d)\}$$

is the set of the (real) zeros of the N-elimination ideal of the following ideal in $N + d$ variables

$$I = \langle t_i(x_j - a(i)_j), \quad i = 1, \ldots, N \text{ and } j = 1, \ldots, d,$$
$$t_1 + \ldots + t_N - 1 \rangle$$

as subset of $k[t_1, \ldots, t_N, x_1, \ldots, x_d]$ where k is a (characteristic zero) field including $a(i)_j$ for all i and j. From Section 2.9 recall that the N-elimination ideal of $I \subset k[t, x]$ is $I \cap k[x]$. The procedure is summarised in the following algorithm.

(i) Write $I \subset k[t_1, \ldots, t_N, x_1, \ldots, x_d]$.

(ii) Find a Gröbner basis G for I with respect to an elimination ordering of the t_i's in $k[t_1, \ldots, t_N, x_1, \ldots, x_d]$.

(iii) The elements of G not containing the t_i's variables are the reduced Gröbner basis for $I \cap k[x_1, \ldots, x_d]$ with respect to the elimination ordering used.

This method can be implemented in any computer algebra package which computes Gröbner bases with respect to an elimination term-ordering. Once we have a Gröbner basis of an ideal with respect to a term-ordering the Gröbner basis with respect to another term-ordering can be computed, for example with the Buchberger algorithm.

Another method to compute design ideals is based on specialized linear algebra techniques for zero-dimensional ideals (see Marinari, Möller and Mora (1993), hence the name M^3 for the corresponding algorithm). A version of this method is implemented in the CoCoA package in the function called IdealOfPoints. It uses the notion of indicator functions (or separators) which we shall come back to in Definition 27.

By Theorem 8, given a term-ordering τ, any polynomial $f \in k[x_1, \ldots, x_d]$ can be decomposed over the design D generated by the Gröbner basis $G = \{g_1, \ldots, g_s\}$ as

$$f = \sum_{i=1}^{s} s_i g_i + r$$

where r is a simpler (with respect to τ) polynomial than f and $f(a(i)) = r(a(i))$ for all design points $a(i) \in D$.

Note once more that we have different Gröbner basis representations of the same ideal corresponding to different term-orderings. Let us give an interpretation in terms of interpolation and with reference to the intersection method to construct the design ideal. For clarity we use the two-dimensional space. Given a set of points $\{(a(i)_1, a(i)_2) : i = 1, \ldots, N\}$ in the plane (x, y) with distinct x values we can always find the unique polynomial of minimum degree $y = p(x)$ through these points. In higher dimension this is no longer true. Unless we fix a term-ordering which, roughly speaking, determines which point to fit first. Gröbner basis theory deals exactly with this problem. Given the design points $\{a(i)_1\}_{i=1,\ldots,N}$ and the observed values $a(i)_2 = p(a(i)_1)$, for $i = 1, \ldots, N$, the remainder of p with respect to the Gröbner basis through the $\{a(i)_1\}_{i=1,\ldots,N}$ points is the minimum polynomial (with respect to the term-ordering) through those points.

Now we have different ways to determine a basis for the design ideal given the variety of the design points. The inverse problem of determining the design points given a finite basis of the design ideal can be solved by computing the reduced Gröbner basis with respect to the term-ordering lex. The obtained triangular polynomial system can be solved by backwards substitution and the solutions are the design points.

3.3 Operations with designs

Theorem 21 (Product of designs) *Let D_1 be a design in k^{d_1} and D_2 a design in k^{d_2}. Let $D_1 \times D_2 = \{(d^1(i), d^2(j)) : d^1(i) \in D_1 \text{ and } d^2(j) \in D_2\}$ be the product design on $k^{d_1+d_2}$. Then, $D_1 \times D_2$ is a design and its ideal is*

$$\text{Ideal}(D_1 \times D_2) = \langle I_1, I_2 \rangle$$

Let τ be a term-ordering over $k[x_1, \ldots, x_{d_1+d_2}]$. Let G_1 be a Gröbner basis for $\text{Ideal}(D_1)$ with respect to the term-ordering obtained by restricting τ to $k[x_1, \ldots, x_{d_1}]$ and let G_2 be the Gröbner basis for $\text{Ideal}(D_2)$ with respect to the term-ordering obtained by restricting τ to $k[x_{d_1}, \ldots, x_{d_1+d_2}]$. Then, $\{g_1, g_2 : g_1 \in G_1 \text{ and } g_2 \in G_2\}$ is a Gröbner basis of $\text{Ideal}(D_1 \times D_2)$ with respect to τ.

The notion of restriction of a term-ordering is intuitive. With the notation of Theorem 21 for x^α, x^β in $k[x_1, \ldots, x_{d_1}]$, $x^\alpha \succ x^\beta$ in the restricted term-ordering if $x^\alpha \succ_\tau x^\beta$ as terms in the larger ring $k[x_1, \ldots, x_{d_1+d_2}]$.

Proof. With the S-polynomial test it can be proved that the set

$$\{g_1, g_2 : g_1 \in G_1 \text{ and } g_2 \in G_2\}$$

is a Gröbner basis. Moreover $D_1 \times D_2$ is a design as the solutions of the system of equations $g_1 = 0, g_2 = 0$ for $g_1 \in G_1$ and $g_1 \in G_1$ are exactly the points in $D_1 \times D_2$. □

Theorem 22 (Restriction) *Let τ be a term-ordering on the monomials of $k[x_1, \ldots, x_d]$. Let D be a design in k^d with $I = \text{Ideal}(D) \subset k[x_1, \ldots, x_d]$ and let J be an ideal in $k[x_1, \ldots, x_d]$. Define the ideal $I + J$ as*

$$I + J = \{f + g : f \in I \text{ and } g \in J\}$$

Let G_1 and G_2 be the τ-Gröbner bases for I and J, respectively. The ideal $I+J$ is the smallest ideal containing I and J and its variety is the restriction of D to $\text{Variety}(J)$, that is $D \cap \text{Variety}(J)$. A basis for $I + J$ is $G = \{g_1, g_2 : g_1 \in G_1 \text{ and } g_2 \in G_2\}$. That is $\text{Variety}(I + J) = D \cap \text{Variety}(J)$.

Proof. See Cox, Little and O'Shea (1997, Proposition 2 Section 4.3). □

The set G in Theorem 22 might not be a Gröbner basis. An example is for $D = 2^2$ as in Example 27 and $J = \langle x_1 - 1, x_1x_2 - 2 \rangle$.

Theorem 23 (Union) *Let D_1 and D_2 be two designs on k^d, let τ be a term-ordering on $k[x_1, \ldots, x_d]$ and G_i the Gröbner basis of $\text{Ideal}(D_i)$ with respect to τ ($i = 1, 2$). The set $D_1 \cup D_2$ is a design and a τ-Gröbner basis (not necessarily reduced) of $\text{Ideal}(D_1 \cup D_2)$ is*

$$\{g_1 g_2 : g_1 \in G_1 \text{ and } g_2 \in G_2\}$$

Proof. See Cox, Little and O'Shea (1997, Theorem 7 Section 4.3). □

Theorem 24 (Image) *Let D be a design on k^d and consider* $\mathrm{Ideal}(D) \subset k[x_1, \ldots, x_d]$. *Let $f_1, \ldots, f_s$ be elements of $k[x_1, \ldots, x_d]$ and define the design $\tilde{D}$, the image of D with respect to $f_1, \ldots, f_s$ as*

$$\tilde{D} = \{f_1, \ldots, f_s\}(D) = \{(f_1(a(j)), \ldots, f_s(a(j))) : a(j) \in D\}$$

If G is a Gröbner basis of $\mathrm{Ideal}(D)$ *with respect to the term-ordering τ then*

$$\mathrm{Ideal}(\tilde{D}) = \langle G, y_i - f_i : i = 1, \ldots, s \rangle \cap k[y_1, \ldots, y_s]$$

admits a Gröbner basis with respect to an elimination ordering of the x-indeterminates whose restriction to the x's is compatible with τ.

Proof. This follows from the elimination theory of Section 2.9. □

3.4 Examples

Example 29 [Full factorial] The 3^3-**full factorial** with three factors at levels $\{-1, 0, 1\}$ corresponds to the variety

$$\mathrm{Variety}\left(x_1^3 - x_1, x_2^3 - x_2, x_3^3 - x_3\right)$$

In general the l^d-full factorial design is represented by the variety

$$\mathrm{Variety}\left(P_1(x_1), \ldots, P_d(x_d)\right)$$

where P_i is the design ideal corresponding to the projection of the full factorial design ideal on the factor i. Thus the degree of P_i is the number of points in the projection over x_i and its roots are the levels of the ith factor. Notice that the symmetry of the design is transfered into the symmetry of the polynomial system defining/interpolating the design points. For the class of models we consider, the identifiability problem is invariant to scaling and shifting of the factors. For example, recoding the levels from $\{-1, 0, 1\}$ to $\{0, 1, 2\}$ in the first variable x_1 corresponds to a shifting of x_1, giving in the above example $x_1^3 + 3x_1^2 + 2x_1$ instead of $x_1^3 - x_1$.

Example 30 The fractional design obtained by the 3^3-full factorial with at least one component zero is the intersection of the varieties $\mathrm{Variety}(x_3^3 - x_3, x_2^3 - x_2, x_1^3 - x_1)$ and $\mathrm{Variety}(x_1 x_2 x_3)$. The result is $\langle x_3^3 - x_3, x_2^3 - x_2, x_1^3 - x_1, x_1 x_2 x_3 \rangle$. Note that the Gröbner basis of both this example and the full factorial designs in Example 29 does not depend on the choice of the term-ordering. Moreover the leading terms of the elements of the Gröbner bases are the same with respect to any term-ordering. We call such bases *total Gröbner bases*. Also check Section 3.9.

Example 31 [3^{4-2}-fractional factorial] The term-ordering becomes essential in the description of the ideal corresponding to a 3^{4-2}-fractional factorial with level set $\{-1, 0, 1\}$. With respect to the `lex` ordering, the Gröbner basis is

$$\begin{cases} x_4^3 - x_4, \\ x_1 + 9/4x_3^2x_4^2 - 3/4x_3^2x_4 - 3/2x_3^2 + 3/4x_3x_4^2 + 3/4x_3x_4 - 1/2x_3 \\ \qquad - 3/2x_4^2 + 1/2x_4 + 1, \\ x_3^3 - x_3, \\ x_2 - 3/2x_3^2x_4 - 3/2x_3x_4^2 + x_3 + x_4 \end{cases}$$

and with respect to `tdeg` it is

$$\begin{cases} x_1x_4 - 1/2x_2x_4 + 1/2x_3x_4 + 1/2x_2 + 1/2x_3, \\ x_2x_3 + x_3^2 - x_2x_4 - x_4^2, \\ x_1x_3 - 1/2x_3^2 + 1/2x_2x_4 - 1/2x_3x_4 + 1/2x_4^2 + 1/2x_2 + 1/2x_4, \\ x_2^2 - x_3^2 + x_2x_4 - x_3x_4, \\ x_1x_2 + 1/2x_3^2 - 1/2x_4^2 + 1/2x_3 + 1/2x_4, x_1^2 + 2x_3^2 - 2x_2x_4 - x_1 - 2, \\ x_3^2x_4 + x_3^2 - x_2x_4 - 2/3x_1 - 1/3x_2 - 2/3x_4 - 2/3, \\ x_3^3 - x_3, x_4^3 - x_4, \\ x_3x_4^2 - x_3^2 + x_2x_4 + 2/3x_1 - 1/3x_2 - 2/3x_3 + 2/3, \\ x_2x_4^2 + x_3^2 - x_2x_4 - 2/3x_1 - 2/3x_2 - 1/3x_3 - 2/3 \end{cases}$$

Example 32 [Echelon designs] A design $D \subset Z_+^d$ is called an *echelon design* if for any design point $(a_1, \ldots, a_d)$ all points of the form $(y_1, \ldots, y_d)$ with $0 \le y_j \le a_j$, for all $j = 1, \ldots, d$ belong to the design D. As an example consider in two dimensions the design

$$D = \{(0,0), (1,0), (2,0), (3,0), (0,1), (1,1)(2,1), (0,2)\}$$

with point pattern

•
• • •
• • • •

A (non-reduced) Gröbner basis for the design ideal with respect to any term-ordering is given by the following five polynomials

$$\begin{cases} x_2(x_2 - 1)(x_2 - 2) \\ x_1x_2(x_2 - 1) \\ x_1(x_1 - 1)x_2(x_2 - 1) \\ x_1(x_1 - 1)(x_1 - 2)x_2 \\ x_1(x_1 - 1)(x_1 - 2)(x_1 - 3) \end{cases}$$

Let now D be a generic echelon design in two dimensions. It is the union of columns of points of the form

$$(0, h), \quad h = 0, \ldots, k_0$$
$$(1, h), \quad h = 0, \ldots, k_1$$
$$\vdots$$
$$(l, h), \quad h = 0, \ldots, k_l$$

where $k_0 \geq k_1 \geq \ldots \geq k_l$. The following polynomials form a Gröbner basis for the design ideal

$$\begin{cases} p_0(x_2) = \prod_{j=0}^{k_0}(x_2 - j) \\ p_1(x_1, x_2) = x_1 \prod_{j=0}^{k_1}(x_2 - j) \\ p_2(x_1, x_2) = x_1(x_1 - 1) \prod_{j=0}^{k_2}(x_2 - j) \\ \vdots \\ p_l(x_1, x_2) = \prod_{j=0}^{l-1}(x_1 - j) \prod_{j=0}^{k_l}(x_2 - j) \\ p_{l+1}(x_1) = \prod_{j=0}^{l}(x_1 - j), \end{cases} \quad (3.1)$$

See Caboara, Pistone, Riccomagno and Wynn (1997) and Robbiano and Rogantin (1998) for an extension to higher dimensions.

3.5 Span of a design

In this section we consider a (single replicate) design D in k^d and are interested in all possible response functions with inputs in D and outputs in an extension field of k.

Example 33 [Continuation of Example 27] Consider all the functions from D to $\mathbb{R}$ where $D = \{\pm 1, \pm 1\} \in \mathbb{Q}^2$. The standard procedure is as follows. Consider the vector $E = [1, x_1, x_2, x_1 x_2]$ and construct the matrix Z with ith row equal to E evaluated at the ith point of D (see also Definition 26 later in this section)

$$Z = \begin{bmatrix} 1 & -1 & -1 & 1 \\ 1 & 1 & -1 & -1 \\ 1 & -1 & 1 & -1 \\ 1 & 1 & 1 & 1 \end{bmatrix}$$

where the columns correspond to $1, x_1, x_2, x_1 x_2$ and the rows to the design points $(-1, -1), (1, -1), (-1, 1), (1, 1)$ in order. Then, interpolate $\theta_0 + \theta_1 x_1 + \theta_2 x_2 + \theta_{12} x_1 x_2$ with $\theta = Z^{-1} y$ where y is a vector of observed values.

For $y = (y_1, y_2, y_3, y_4)$ we obtain

$$\hat{\theta}_1 = \frac{1}{4}(y_1 + y_2 + y_3 + y_4)$$
$$\hat{\theta}_2 = \frac{1}{4}(-y_1 + y_2 - y_3 + y_4)$$
$$\hat{\theta}_3 = \frac{1}{4}(-y_1 - y_2 + y_3 + y_4)$$
$$\hat{\theta}_4 = \frac{1}{4}(y_1 - y_2 - y_3 + y_4)$$

In the previous example the vector E was a known regression vector. Given a design with N points and N functions $f_i : D \to \mathbb{R}$ the same procedure applies if the regression vector $E = (f_j : j = 1, \ldots, N)$ is such that the vectors $E(a)$, $a \in D$ are linearly independent.

Next, we give a method to find suitable vectors E given a design D. A version of Theorem 15 of Chapter 2 is at the heart of the theory. Let D be a design and τ a term-ordering, a monomial basis of the set of polynomial functions over D is

$$\mathrm{Est}_\tau(D) = \{\, x^\alpha : x^\alpha \text{ is not divisible by any of the leading terms}$$
$$\text{of the elements of the Gröbner basis of Ideal}(D)\,\}$$
$$= \{x^\alpha : x^\alpha \notin \langle \mathrm{LT}(g) : g \in \mathrm{Ideal}(D)\rangle\}$$

See also Definition 29.

Let us notice that $Est_\tau(D)$ is of echelon form. Some literature refers to this by saying that $Est_\tau(D)$ is an *order ideal*, that is, if $x^\alpha \in Est_\tau(D)$ and x^β divides x^α, then $x^\beta \in Est_\tau(D)$.

In general, different term-orderings give different $Est_\tau(D)$. All are order ideals and have the same number of elements, as they are different bases of the same vector space, namely the quotient space $k[x]/\mathrm{Ideal}(D)$. Most importantly, the following theorem holds.

Theorem 25 *The set $Est_\tau(D)$ has as many elements as there are design points.*

Proof. This follows from Theorem 26 below. □

We see in Section 3.6 that this has consequences in modeling.

We can index $Est_\tau(D)$ over a list L as

$$\{x^\alpha : \alpha \in L\}$$

where L is the set of exponents of the elements in $Est_\tau(D)$.

Definition 26 *Let τ be a term-ordering and let us consider an ordering over the design points $D = \{a(i) \in k^d : i = 1, \ldots, N\}$. Let L be the set of exponents of $Est_\tau(D)$. We call* design matrix *the following matrix*

$$Z = [a(i)^\alpha]_{i=1,\ldots,N;\alpha \in L}$$

Theorem 26

1. Z is non-singular.
2. Let e_i be the d-dimensional vector with components 0 except in position i where it has value 1. For all $i = 1, \ldots, d$ there exists a vector $c(i) \in k^d$ such that $Zc(i) = e_i$ and the polynomial $\sum_{\alpha \in L} c(i)_\alpha x^\alpha$ interpolates the indicator function of the design point $a(i)$. That is

$$\sum_{\alpha \in L} c(i)_\alpha x^\alpha = \begin{cases} 1 & x = a(i) \\ 0 & x \neq a(i) \end{cases} \text{ and } x \in D$$

3. If $f : D \to \mathcal{K}$ is a response mapping and $y = (f(a(1)), \ldots, f(a(N))) \in \mathcal{K}$ is the vector of responses, then

$$f(x) = \sum_{\alpha \in L} c_\alpha x^\alpha \qquad (3.2)$$

where the vector of coefficients $c = [c_\alpha : \alpha \in L] = Z^{-1}y$ and $\mathcal{K}$ is an extension of k.

Proof. The proof is a simple check and follows immediately from the definitions and the theory in Chapter 2. □

The usual estimator of the parameters $(Z^t Z)^{-1} Z^t y$ simplifies here to $Z^{-1}y$ as Z is square full rank. In statistical terms the model is saturated. The representation (3.2) of experimental results as an interpolatory polynomial of minimal complexity (with respect to τ) is unusual and is one of the key ideas of the book.

Definition 27 *Given a design $D = \{a(i) \in k^d : i = 1, \ldots, N\}$ we call the indicator polynomial (or interpolatory polynomial) of $a(i)$ a polynomial $f \in k[x_1, \ldots, x_d]$ such that*

$$f(x) = \begin{cases} 1 & x = a(i) \\ 0 & x \neq a(i) \end{cases} \text{ and } x \in D$$

Example 34 The polynomial defined in Theorem 26 Item 2 is an indicator polynomial of D of minimum complexity with respect to the chosen term-ordering τ.

Example 35 Consider the three points $a(1) = (0,0), a(2) = (0,1)$ and $a(3) = (1,0)$ and the corresponding echelon design $D = \{a(i) \in \mathbb{C}^2 : i = 1, 2, 3\}$ over the complex field. The design ideal is $\text{Ideal}(D) = \langle x_1^2 - x_1, x_2^2 - x_2, x_1 x_2 \rangle$ and a basis of the vector space $\mathbb{C}[x_1, x_2]/\text{Ideal}(D)$ is $\{1, x_2, x_1\}$ with respect to any term-ordering. The elements of the list L are $(0,0), (0,1), (1,0)$ and the design matrix Z is

$$Z = \begin{bmatrix} 1 & 0 & 0 \\ 1 & 1 & 0 \\ 1 & 0 & 1 \end{bmatrix}$$

where the columns are labelled 1, x_2 and x_1 left to right and the rows are labelled $(0,0)$, $(0,1)$ and $(1,0)$ top to bottom. The inverse of Z is

$$Z^{-1} = \begin{bmatrix} 1 & 0 & 0 \\ -1 & 1 & 0 \\ -1 & 0 & 1 \end{bmatrix}$$

Now the ith row of the vector

$$(Z^{-1})^t \begin{bmatrix} 1 \\ x_2 \\ x_1 \end{bmatrix} = \begin{bmatrix} 1 - x_2 - x_1 \\ x_2 \\ x_1 \end{bmatrix}$$

is the interpolatory polynomial of the design point $a(i)$. Now let us suppose we have the vector $y = (y_1, y_2, y_3) \in \mathbb{C}^3$ of given values associated to the design points. The equation

$$\sum_{i=1}^{3} y_i \sum_{\alpha \in L} c_\alpha x^\alpha = \sum_{\alpha \in L} x^\alpha \sum_{i=1}^{3} y_i c_\alpha = y^t (Z^{-1})^t [x^\alpha]_{\alpha \in L}$$

$$= (y_1, y_2, y_3) \begin{bmatrix} 1 & -1 & -1 \\ 0 & 0 & 1 \\ 0 & 1 & 0 \end{bmatrix} \begin{bmatrix} 1 \\ x_2 \\ x_1 \end{bmatrix}$$

$$= y_1 + (y_2 - y_1)x_2 + (y_3 - y_1)x_1$$

gives a polynomial whose value at $a(i)$ is y_i.

3.6 Models and identifiability: quotients

The results of the previous section can be seen as providing statements about identifiability.

Theorem 27 *Let D be a design and τ a term-ordering. The model*

$$\sum_{x^\alpha \in \mathrm{Est}_\tau(D)} \theta_\alpha x^\alpha \qquad (3.3)$$

is unambiguously identifiable since at the design points $a(i)$ and for the observed values y_i, for all $a(i) \in D$, the linear system of equations

$$y_i = \sum_{x^\alpha \in \mathrm{Est}_\tau(D)} \theta_\alpha x^\alpha(a(i))$$

has one and only one solution with respect to θ_α. That is D identifies the model in Equation (3.3).

Proof. From the arguments in Section 3.5, the above is a non-degenerate system of linear equations in θ_α's with as many unknowns as equations. □

Corollary 2 *With an N point-design we can identify (up to) N distinct terms.*

Proof. By Theorem 15 we have that we can always identify the same number of terms whatever term-ordering we use, and by the elimination theory it follows that this number is N. □

Note that while the dimension is independent of the ordering, the elements of $\text{Est}_\tau(D)$ strongly depend on the chosen term-ordering and thus we have in general a whole range of identifiable saturated models (see also Section 3.9). This fact could be used to influence the model structure. For example, when main effects are favored, then a term-ordering that respects the total degree of terms, such as the tdeg ordering, may be used. Another example is when one effect dominates all the others. Then, a lexicographic ordering may be appropriate (see Section 3.13).

Also Est is an order ideal, and this reflects one common practice in modeling of including all the factors of a present interaction, that is, all lower-order terms which divide a given term. McCullagh and Nelder (1983) speak of *functional marginality*. Also the term *hierarchical model* is appropriate for models with the order ideal property.

Corollary 3 *Any element of* $\text{Est}_\tau(D)$ *is the representative of an equivalence class (congruent to the design ideal).*

Proof. This follows from the definition of $\text{Est}_\tau(D)$. □

In particular, in the model in Equation (3.3), a monomial x^α, $\alpha \in L$ can be substituted with another element in the equivalence class of x^α, not necessarily a monomial (see below).

Note that once we have determined the Gröbner basis with respect to a given term-ordering, the set of identifiable terms is automatically computed without reference to the term-ordering chosen, since it only depends on the properties of monomial multiplication.

With the above notation, the concept of algebraic identifiability is summarized by the following mapping

$$\begin{aligned} \mathcal{I}_{D,\tau} : k(\theta_0, \ldots, \theta_p)[x] &\longrightarrow k(\theta_0, \ldots, \theta_p)[x]/\text{Ideal}(D) \\ f &\longmapsto \text{Rem}(f, G) \end{aligned}$$

where we stress the presence of parameters in the coefficient field, see Caboara and Riccomagno (1998). Note that $\mathcal{I}_{D,\tau}$ is not the congruence modulo $\text{Ideal}(D)$. Indeed it concentrates on the vector-space structure of the quotient ideal, and the defining operation is the division with respect to the Gröbner basis G.

3.7 Confounding of models

It is useful to highlight the interpolation aspect of Theorem 27. As noted, since the number of estimable terms is exactly the sample size, we obtain *exact interpolation* when we fit the linear model composed exactly of all identifiable terms. Any submodel of terms from $\text{Est}_\tau(D)$ is also full rank and can provide a candidate model. Such submodels will have an important place in this book, particularly in Chapter 6.

It is important to emphasize that we start with the whole polynomial set, $k[x]$. Given the term-ordering τ and the design D, let $G = \{f_1, \ldots, f_v\}$ be the τ-Gröbner basis of Ideal(D). The vector space $k[x]/\text{Ideal}(D)$ is the set of classes of remainders of the polynomials of $k[x]$ with respect to division by G. Thus, for $f \in k[x]$, the equivalence class of f in $k[x]/\text{Ideal}(D)$ is

$$\{g \in k[x] : f - g \in \text{Ideal}(D)\}$$

A representative of the equivalence class of f is $\text{Rem}(f, G)$ where G is a Gröbner basis of Ideal(D). One interpretation is that the equivalence class of a certain polynomial f gives all the polynomials that interpolate the values of f at the design points.

Given a model f, a term-ordering τ and a design D, a model identifiable by D and confounded with f is $\text{Rem}(f, G)$. In particular, if $\text{Rem}(f, G)$ is f then f is identifiable by D. Thus the problem of checking whether a model is identifiable by a design consists of computing and checking a remainder. This operation can easily be carried out in CoCoA and Maple using the NF and normalf commands, respectively. This holds because the division algorithm operates linearly on the coefficients/parameters of f and thus on the parameters of the model in such a way that if f is identifiable with respect to a certain term-ordering, then it is identifiable with respect to any term-ordering. Identifiability is a property of designs and models and does not depend on the term-ordering.

Corollary 4 *If a model M is identifiable by D with respect to a term-ordering τ according to Theorem 27, then M is identifiable by D with respect to any other term-ordering σ.*

Proof. A non-singular transformation transforms Est_σ to Est_τ. □

Faugère, Gianni, Lazard and Mora (1993) compute efficiently Gröbner bases for design ideals with respect to different term-orderings via the notion of a Gröbner walk.

Example 36 Consider the 2^2 design with levels 0, 1. With respect to any term-ordering the reduced Gröbner basis is $\{x_1^2 - x_1, x_2^2 - x_2\}$ and $\text{Est}(D) = \{1, x_1, x_2, x_1x_2\}$. The full quadratic model

$$f(x_1, x_2) = \theta_1 x_1^2 + \theta_2 x_2^2 + \theta_3 x_1 x_2$$

is identifiable with D as $\text{Rem}(f, G) = \theta_1 x_1 + \theta_2 x_2 + \theta_3 x_1 x_2$. While the full cubic model

$$h(x_1, x_2) = \theta_1 x_1^3 + \theta_2 x_2^3 + \theta_3 x_1^2 x_2 + \theta_4 x_1 x_2^2$$

is not identifiable as $\text{Rem}(h, G) = \theta_1 x_1 + \theta_2 x_2 + (\theta_3 + \theta_4) x_1 x_2$ and θ_3 and θ_4 are confounded. Notice, however, that the design $D_1 = \{-1, +1\}^2$ has the same Est set as the previous one, but a different Gröbner basis $G_1 = \{x_1^2 - 1, x_2^2 - 1\}$. In particular, we have $\text{Rem}(f, G_1) = \theta_1 + \theta_2 + \theta_3 x_1 x_2$, that is, the full quadratic model is not identifiable by D_1.

Each of the equivalence classes in $k[x]/\text{Ideal}(D)$ can be seen as an aliasing class in the sense that only one term from each class can be part of the same identifiable model. See Holliday, Pistone, Riccomagno and Wynn (1999). That is, any residual class of $k[x]/\text{Ideal}(D)$ is an infinite family of models that are not distinguishable by the design.

From the above discussion we have that the following definition is well posed.

Definition 28 *Two models, f and g, are* confounded *(aliased) under the design D if and only if $f - g$ belongs to the design ideal $\text{Ideal}(D)$, equivalently $\mathcal{I}_{D,\tau}(f) = \mathcal{I}_{D,\tau}(g)$ for any term-ordering.*

Given the design D, we select identifiable models as follows. As regression vectors, choose any subset of a basis of the vector space $k[x]/\text{Ideal}(D)$. Thus algebraic estimability becomes the following mapping

$$\mathcal{I}_{D,\tau} : \mathbb{Q}(\theta_0, \ldots, \theta_p)[x_1, \ldots, x_d] \times \mathbb{Q}^N \longrightarrow \mathbb{Q}[x_1, \ldots, x_d]/\text{Ideal}(D)$$
$$(f, y) \longmapsto X(x)(X^t X)^{-1} X^t y$$

where N is the design size, $X(x)$ is the regression vector extracted from $\text{Rem}(f, G)$ and X is the "design matrix" for $X(x)$ and D. Note that X is a sub-matrix of the design matrix in Definition 26. We consciously used the same notation for algebraic identifiability and estimability to stress that they often correspond both in theory and applications.

3.8 Further examples

Example 37 [Continuation of Example 29] The set of terms identifiable by the 3^3-full factorial with respect to the `tdeg` term-ordering is the following

Est = { $x_1^2 x_2^2 x_3^2$,
$x_1^2 x_2^2 x_3$, $x_1^2 x_2 x_3^2$, $x_1 x_2^2 x_3^2$,
$x_1^2 x_2^2$, $x_1^2 x_2 x_3$, $x_1 x_2^2 x_3$, $x_1^2 x_3^2$, $x_1 x_2 x_3^2$, $x_2^2 x_3^2$,
$x_1^2 x_2$, $x_1 x_2^2$, $x_1^2 x_3$, $x_1 x_2 x_3$, $x_2^2 x_3$, $x_1 x_3^2, x_2 x_3^2$,
x_1^2, $x_1 x_2$, x_2^2, $x_1 x_3$, $x_2 x_3$, x_3^2,
x_1, x_2, x_3, 1 }

We have the well known result that the largest model we can identify with the 3^3-full factorial design is the "full product" model. This is the case for all term-orderings, since the Gröbner basis $\{x_1^3 - x_1, x_2^3 - x_2, x_3^3 - x_3\}$ is a total Gröbner basis.

Example 38 [Continuation of Example 30] The subset of 3^3-full factorial

FURTHER EXAMPLES

for which at least one factor is zero gives the following subset of identifiable terms with respect to any term-ordering

$$\text{Est} = \{ \; x_1^2 x_2^2, \quad x_1^2 x_3^2, \quad x_2^2 x_3^2, $$
$$x_1^2 x_2, \quad x_1 x_2^2, \quad x_1^2 x_3, \quad x_2^2 x_3, \quad x_1 x_3^2, \quad x_2 x_3^2 $$
$$x_1^2, \quad x_1 x_2, \quad x_2^2, \quad x_1 x_3, \quad x_2 x_3, \quad x_3^2, $$
$$x_1, \quad x_2, \quad x_3, \quad 1 \; \}$$

Example 39 With the notation of Theorem 21 and with respect to τ, the Est set for a product design $D_1 \times D_2$ is the product of the Est sets for the single designs and similarly for multiple products. This follows directly from the structure of the Gröbner basis. Restrictions and union of designs are discussed in Section 3.12.

Example 40 [Continuation of Example 31] For the 3^{4-2} fractional full factorial design with the tdeg ordering with

$$x_1 \succ x_2 \succ x_3 \succ x_4$$

we have

$$\text{Est}_{\text{tdeg}}(3^{4-2}) = \{ \; x_3^2, \quad x_2 x_4, \quad x_3 x_4, \quad x_4^2, $$
$$x_1, \quad x_2, \quad x_3, \quad x_4, \quad 1 \; \}$$

and with the lex ordering for $x_1 \succ x_2 \succ x_3 \succ x_4$

$$\text{Est}_{\text{lex}}(3^{4-2}) = \{ \; x_3^2 x_4^2, \quad x_3^2 x_4, \quad x_3^2, $$
$$x_3 x_4^2, \quad x_3 x_4, \quad x_3, $$
$$x_4^2, \quad x_4, \quad 1 \; \}$$

Two factors, x_1 and x_2, are not in the above list of identifiable terms as can be expected from the property of lexicographic ordering. Note here the fact that the cardinality of these last two Est sets is equal to 9, the number of design points.

Example 41 Consider the quadratic model in one variable and the three-point design $\{a(1), a(2), a(3)\}$. Thus the design ideal is generated by $(x - a(1))(x - a(2))(x - a(3))$ and the model is $\theta_0 + \theta_1 x + \theta_2 x^2$. With respect to the only term-ordering in one dimension the remainder is $\theta_0 + \theta_1 x + \theta_2 x^2$. As expected all three parameters are identifiable.

Instead in the cubic model the linear and cubic effects are aliased as the remainder of the cubic model equation

$$\theta_0 + \theta_1 x + \theta_2 x^2 + \theta_3 x^3$$

by the design ideal is

$$\theta_0 + a(1)a(2)a(3)\theta_3 + (\theta_1 - \theta_3 a(2)a(3) - \theta_3 a(1)a(3) - \theta_3 a(1)a(2)) x$$
$$+ (\theta_3 + \theta_3 a(3) + \theta_3 a(2) + \theta_3 a(1)) x^2$$

Example 42 An early example of the theory, Pistone and Wynn (1996), gives a useful insight on the interpolation issue. Consider three points in generic position in the plane,

$$(a(1)_1, a(1)_2), (a(2)_1, a(2)_2), (a(3)_1, a(3)_2)$$

The reduced Gröbner basis with respect to the term-ordering $\texttt{lex}\,(x_1 \succ x_2)$ is composed of the following two polynomials

$$\begin{aligned}
g1 = &-a(3)_2 a(1)_1 a(2)_2^2 + a(3)_2^2 a(1)_1 a(2)_2 + a(3)_2 a(2)_1 a(1)_2^2 \\
&- a(3)_2^2 a(2)_1 a(1)_2 + a(3)_1 a(2)_2^2 a(1)_2 - a(3)_1 a(2)_2 a(1)_2^2 \\
&+ \left(a(1)_1 a(2)_2^2 - a(1)_1 a(3)_2^2 + a(2)_1 a(3)_2^2 - a(2)_1 a(1)_2^2\right. \\
&\left. + a(3)_1 a(1)_2^2 - a(3)_1 a(2)_2^2\right) x_2 \\
&+ \left(-a(1)_1 a(2)_2 + a(1)_1 a(3)_2 - a(2)_1 a(3)_2 + a(3)_1 a(2)_2\right. \\
&\left. + a(2)_1 a(1)_2 - a(3)_1 a(1)_2\right) x_2^2 \\
&+ \left(a(2)_2^2 a(3)_2 - a(2)_2 a(3)_2^2 - a(3)_2 a(1)_2^2\right. \\
&\left. + a(3)_2^2 a(1)_2 - a(2)_2^2 a(1)_2 + a(2)_2 a(1)_2^2\right) x_1, \\
g2 = &-a(3)_2 a(2)_2 a(1)_2 + \left(a(2)_2 a(1)_2 + a(3)_2 a(1)_2 + a(2)_2 a(3)_2\right) x_2 \\
&+ \left(-a(1)_2 - a(2)_2 - a(3)_2\right) x_2^2 + x_2^3
\end{aligned}$$

The leading terms and their coefficients are

LT	Coefficient
x_1	$a(2)_2^2 a(3)_2 - a(2)_2 a(3)_2^2 - a(3)_2 a(1)_2^2$
	$+ a(3)_2^2 a(1)_2 - a(2)_2^2 a(1)_2 + a(2)_2 a(1)_2^2$
x_2^3	1

The set of identifiable terms is $1, x_2, x_2^2$. We notice that the coefficient of x_1 is zero if and only if at least two of the design points have the same x_2 value. Repeating the above procedure with respect to $\texttt{lex}$ and with the constraint $a(1)_2 = a(2)_2$, the set of leading terms becomes $x_1 x_2, x_1^2$. Thus the identifiable terms are $1, x_2, x_1$. The same result is obtained when the calculations are carried out with respect to the $\texttt{tdeg}$ ordering. This is an example of the connection between the structure of a design and the set of identifiable terms. A major area of future work is to analyze the link between the geometry structure of designs and the set of identifiable terms returned by the above procedure for a fixed term-ordering. Some initial results are contained in the study of fans in Section 3.9.

Example 43 [Continuation of Example 37] In Section 3.4 we have seen that with the 3^3-full factorial design, we can estimate the standard quadratic model. Next, we show the confounding structure for a standard cubic model under a 3^3-full factorial with obvious notation

$$(\theta_{2,1,0} + \theta_{2,3,0})\, x_1^2\, x_2 + \theta_{2,0,2}\, x_1^2\, x_3^2 + \theta_{2,0,0}\, x_1^2 + \theta_{2,2,0}\, x_1^2\, x_2^2$$
$$+ \theta_{2,2,2}\, x_1^2\, x_2^2\, x_3^2 + (\theta_{1,2,0} + \theta_{3,2,0})\, x_1\, x_2^2$$
$$+ (\theta_{2,0,1} + \theta_{2,0,3})\, x_1^2\, x_3 + (\theta_{0,2,1} + \theta_{0,2,3})\, x_2^2\, x_3$$
$$+ (\theta_{1,0,2} + \theta_{3,0,2})\, x_1\, x_3^2$$
$$+ (\theta_{1,1,0} + \theta_{3,1,0} + \theta_{1,3,0} + \theta_{3,3,0})\, x_1\, x_2$$
$$+ (\theta_{1,0,1} + \theta_{3,0,1} + \theta_{1,0,3} + \theta_{3,0,3})\, x_1\, x_3$$
$$+ (\theta_{0,1,1} + \theta_{0,3,1} + \theta_{0,1,3} + \theta_{0,3,3})\, x_2\, x_3$$
$$+ (\theta_{0,1,2} + \theta_{0,3,2})\, x_2\, x_3^2 + (\theta_{0,1,0} + \theta_{0,3,0})\, x_2$$
$$+ (\theta_{1,1,2} + \theta_{3,1,2} + \theta_{1,3,2} + \theta_{3,3,2})\, x_1\, x_2\, x_3^2 + (\theta_{1,1,1} + \theta_{3,1,1}$$
$$+ \theta_{1,3,1} + \theta_{3,3,1} + \theta_{1,1,3} + \theta_{3,1,3} + \theta_{1,3,3} + \theta_{3,3,3})\, x_1\, x_2\, x_3$$
$$+ (\theta_{2,1,2} + \theta_{2,3,2})\, x_1^2\, x_2\, x_3^2 + (\theta_{1,2,2} + \theta_{3,2,2})\, x_1\, x_2^2\, x_3^2$$
$$+ (\theta_{2,1,1} + \theta_{2,3,1} + \theta_{2,1,3} + \theta_{2,3,3})\, x_1^2\, x_2\, x_3$$
$$+ (\theta_{1,2,1} + \theta_{3,2,1} + \theta_{1,2,3} + \theta_{3,2,3})\, x_1\, x_2^2\, x_3$$
$$+ (\theta_{2,2,1} + \theta_{2,2,3})\, x_1^2\, x_2^2\, x_3 + (\theta_{0,0,1} + \theta_{0,0,3})\, x_3$$
$$+ (\theta_{1,0,0} + \theta_{3,0,0})\, x_1 + 2\,\theta_{0,0,0} + \theta_{0,2,2}\, x_2^2\, x_3^2 + \theta_{0,2,0}\, x_2^2$$
$$+ \theta_{0,0,2}\, x_3^2$$

where $\theta_{i,j,k}$, for $i,j,k = 0,1,2$ is the parameter of the term $x_1^i x_2^j x_3^k$ in the cubic model. Of the 64 parameters of the cubic model only the seven coefficient terms involving only second-order powers and the constant are fully identifiable. Notice that the coefficients of the other elements of Est are linear combinations of the model parameters. As previously noticed, this is always the case, since the division operates linearly on the coefficients of the dividend.

Example 44 [Example 30 continued] Another example of parameter confounding is the standard quadratic model in 3 dimensions and the subset of 3^3-full factorial with at least one zero-component.

$$\theta_{0,1,2}\, x_2\, x_3^2 + \theta_{2,0,0}\, x_1^2 + \theta_{0,2,0}\, x_2^2 + \theta_{1,0,2}\, x_1\, x_3^2 + \theta_{0,0,1}\, x_3$$
$$+ \theta_{0,0,2}\, x_3^2 + \theta_{1,1,0}\, x_1\, x_2 + \theta_{1,0,0}\, x_1 + \theta_{0,1,0}\, x_2 + \theta_{2,1,0}\, x_1^2\, x_2$$
$$+ \theta_{0,2,2}\, x_2^2\, x_3^2 + \theta_{0,1,1}\, x_2\, x_3 + 2\,\theta_{0,0,0} + \theta_{0,2,1}\, x_2^2\, x_3 + \theta_{1,0,1}\, x_1\, x_3$$
$$+ \theta_{1,2,0}\, x_1\, x_2^2 + \theta_{2,0,1}\, x_1^2\, x_3 + \theta_{2,0,2}\, x_1^2\, x_3^2 + \theta_{2,2,0}\, x_1^2\, x_2^2$$

and thus the product model is identifiable with a 3^3-full factorial design with at least one zero component.

3.9 The fan of an experimental design

In this section we define a large class of models with an order ideal structure and identifiable by a given design. The idea of a fan of a design is introduced in experimental design by Caboara, Pistone, Riccomagno and Wynn (1997). The algebraic notion of fan of an ideal goes back to Mora and Robbiano (1988). We start with some definitions.

Definition 29 *Let G be a Gröbner basis with respect to a term-ordering τ. The monomial set*

$$\mathrm{Init}_\tau(G) = \mathrm{Ideal}(\mathrm{LT}_\tau(g) : g \in G)$$

is the initial ideal *of G with respect to τ.*

Note that, by the definition of a Gröbner basis, the following holds

$$\mathrm{Init}_\tau(G) = \mathrm{Ideal}(\mathrm{LT}_\tau(g) : g \in I)$$

where I is the ideal generated by G. Thus we also write $\mathrm{Init}_\tau(I)$. Note that the complementary set of $\mathrm{Init}_\tau(G)$ is an order ideal, and if G is a Gröbner basis for a design ideal, then such complementary set is an Est set.

Theorem 28 *Every ideal $I \subset k[x]$ has only finitely many distinct initial ideals, equivalently order ideals.*

Proof. See, for example, Sturmfels (1996). □

Corollary 5 *Given an ideal I (in particular, a design D), the following defines an equivalence relation on the set of all term-orderings. The term-orderings τ_1 and τ_2 are equivalent with respect to I if and only if they have the same initial ideal*

$$\mathrm{Init}_{\tau_1}(I) = \mathrm{Ideal}(\mathrm{LT}_{\tau_1}(g) : g \in G_{\tau_1}) = \mathrm{Ideal}(\mathrm{LT}_{\tau_2}(g) : g \in G_{\tau_2}) = \mathrm{Init}_{\tau_2}(I)$$

where G_{τ_j} is the Gröbner basis of I with respect to τ_j, $(j = 1, 2)$.

Proof. It follows directly from Theorem 28. □

Note that to each equivalence class one can associate an order ideal, namely the complementary set. In particular, when I is a design ideal $\mathrm{Ideal}(D)$, such complementary set is an Est set. This leads to the definition of *fan* of a polynomial ideal.

Definition 30 *Given the polynomial ideal I, the partition on the set of term-orderings induced by the equivalence relation in Corollary 5 is called the* (algebraic) fan *of I. Each element in the fan is called a* leaf.

Each leaf of a fan is represented by an initial ideal. Equivalently it is represented by the complementary set of the initial ideal. Now the notion of *fan of a design* is well defined.

Definition 31 *Let D be a design. The fan of D is the set of all $\text{Est}_\tau(D)$ as the term-ordering τ varies over all term-orderings.*

The following example by Caboara and Robbiano (1997) shows that identifiability via Gröbner basis is not exhaustive in the sense that, given a design D, there are models identifiable by D, namely whose design matrix is full rank, that are not retrievable by the methods in this book.

Example 45 The design $D = \{(0,0),(0,-1),(1,0),(1,1),(-1,1)\}$ identifies the model corresponding to the order ideal $\{1, x_1, x_2, x_1^2, x_2^2\}$, but such ideal does not belong to the fan of D. The fan of D consists of the two leaves $\{1, x_1, x_1^2, x_2, x_1 x_2\}$ and $\{1, x_2, x_2^2, x_1, x_1 x_2\}$.

In the following example we show how a new concept of model confounding can overcome the above problem.

Example 46 For the term-ordering $\texttt{tdeg}\,(x_1 \succ x_2)$, the Gröbner basis for the design in the previous example is

$$\begin{cases} x_1^2 + x_1 x_2 - \tfrac{1}{2} x_2^2 - x_1 - \tfrac{1}{2} x_2 \\ x_2^3 - x_2 \\ x_1 x_2^2 - x_1 x_2 \end{cases}$$

The remainder of the polynomial model

$$f(x) = \theta_0 + \theta_1 x_1 + \theta_2 x_1^2 + \theta_3 x_2 + \theta_4 x_2^2$$

is

$$\theta_0 - \theta_2 x_1 x_2 + (\theta_1 + \theta_2) x_1 + (\theta_3 + \tfrac{1}{2}\theta_2) x_2 + (\theta_4 + \tfrac{1}{2}\theta_2) x_2^2$$

There is an invertible linear relationship between the coefficients of the remainder and the θ's. This is always the case because the division algorithm operates linearly on the coefficients of f. Thus, according to Definition 28, f is identifiable.

Example 47 Let D be the star composite design with central point in d dimensions. To fix notation, assume that the central point is $0 = (0, \ldots, 0)$, the levels of the 2^d full factorial part are ± 1 and that the arms on each axis are at levels ± 2. Then, the fan of D has d leaves. One leaf (with respect to any term-ordering such that $x_i \succ x_d$ for all $i = 1, \ldots, d-1$) is

$$L = \{\quad 1$$
$$\phantom{L = \{\quad} x_i^2 \quad \text{(for all } i = 1, \ldots, d)$$
$$\phantom{L = \{\quad} x_1^4$$
$$\phantom{L = \{\quad} x_i x_1^2 \quad \text{(for all } i = 1, \ldots, d)$$
$$\phantom{L = \{\quad} \prod_{i \in I} x_i \quad \text{(for all } I \text{ with } N \text{ elements and } I \subset \{1, \ldots, d\}$$
$$\phantom{L = \{\quad \prod_{i \in I} x_i \quad \text{(for }} \text{and } N = 1, \ldots, d)\quad \}$$

The other leaves are obtained by permutation of the variables. For the proof we refer to Caboara, Pistone, Riccomagno and Wynn (1997).

3.9.1 Computation of fans

The computation of the fan of a design (and of an ideal in general) is very expensive as it involves the computation of many Gröbner bases. It is known that, in the worst case, the computation of Gröbner bases has double exponential cost. Ideally one would like to input all the information available on the term-ordering before starting the computation, that is to define a pre-ordering instead of a term-ordering.

The algorithm to calculate fans of ideals receives as input a basis of the design ideal. At each step it chooses the possible leading terms compatible with the known ordering information, applies the S-polynomial test to check whether a set of polynomials is a Gröbner basis with respect to some term-ordering, and keeps iterating to create new leaves of the fan. When the S-polynomial test is positive over one leaf, it returns the Gröbner basis associated with that leaf and the conditions which the term-ordering of that leaf must satisfy. This algorithm was first introduced in Mora and Robbiano (1988). The usual improvements to the Buchberger algorithm for reduced Gröbner bases can be applied. We show the details with an example.

Example 48 Consider the design $D = \{(0,0), (1,2), (2,1)\}$ and impose the condition $x_1 \succ x_2$ on the term-ordering. The design D is the set of solution of the following system of polynomial equations

$$f = x_2^3 - 3x_2^2 + 2x_2$$
$$g = x_1 + 3/2x_2^2 - 7/2x_2$$

The possible leading terms of g (compatible with $x_1 \succ x_2$) are x_1 and x_2^2, and for f we have only x_2^3. We create two leaves characterised by the conditions $x_1 \succ x_2^2$ and $x_2^2 \succ x_1$, respectively. The S-polynomials are

$$\text{S-poly}(f,g) = -3x_2^2 x_1 + 2x_1 x_2 - 3/2x_2^5 + 7/2x_1^4 \qquad \text{for } x_1 \succ x_2^2$$
$$\text{S-poly(f,g)} = -\frac{2}{3}x_2^2 + 2x_2 - \frac{2}{3}x_2 x_1 \qquad \text{for } x_2^2 \succ x_1$$

Their remainders with respect to f and g are

$$p = \text{Rem}(\text{S-poly}(f,g), \{f,g\}) = 0 \qquad \text{for } x_1 \succ x_2^2$$
$$h = \text{Rem}(\text{S-poly}(f,g), \{f,g\}) = -\frac{2}{3}x_1 x_2 + \frac{4}{9}x_1 + \frac{4}{9}x_2 \qquad \text{for } x_2^2 \succ x_1$$

Since $p = 0$, by the S-polynomial test we have that, for all the orderings such that $x_1 \succ x_2^2$, the set $\{f,g\}$ is a (reduced) Gröbner basis, which gives $\{1, x_2, x_2^2\}$ as the estimable set.

We have to continue the calculation for the orderings such that $x_2^2 \succ x_1$. The new generating set is $\{f, g, h\}$ and the only possible leading term of h

is $x_1 x_2$. Thus

$$\text{S-poly}(f, h) = -7/3 x_1 x_2^2 + 2 x_1 x_2 + 2/3 x_2^3$$
$$\text{S-poly}(g, h) = \frac{2}{3}(x_1^2 + x_2^2) - \frac{5}{3} x_1 x_2$$

and

$$l = \text{Rem}(\text{S-poly}(f, h), \{f, g, h\}) = -\frac{14}{9} x_1^2 + \frac{98}{27} x_1 - \frac{28}{27} x_2$$
$$m = \text{Rem}(\text{S-poly}(g, h), \{f, g, h\}) = \frac{2}{3} x_1^2 - \frac{14}{9} x_1 + \frac{4}{9} x_2$$

Because of the condition $x_1 \succ x_2$ on the ordering, the only possible leading term of l and g is x_1^2. The S-polynomial test shows that for the term-orderings such that $x_2^2 \succ x_1$ and $x_1 \succ x_2$ the set $\{f, g, h, l, m\}$ is a Gröbner basis. The estimable set is $\{1, x_1, x_2\}$. In conclusion the fan of the design D with the constrained $x_1 \succ x_2$ is $\{\{1, x_2, x_2^2\}, \{1, x_1, x_2\}\}$.

If no condition on the ordering is imposed, the above algorithm returns the fan of the ideal given as input. Alternatively, to compute the fan one could use the so-called Gröbner walk technique. A Gröbner basis is computed with respect to some ordering, usually tdeg, and then from such basis the bases for the other leaves are computed in a linear time. See Collart, Kalkbrener and Mall (1997) and Faugère, Gianni, Lazard and Mora (1993).

3.10 Minimal and maximal fan designs

From Definition 31 it follows that designs can be classified according to the number of leaves in their fan.

Definition 32 *A design is called* minimal fan *if its fan has only one leaf.*
Note that a minimal fan design could actually identify other saturated models with an order ideal structure, but such models would not be retrieved with the Gröbner basis method directly.

Definition 33 *A design $D \subset Z_+^d$ is called a* generalised echelon design *if for any design point $(a_1, \ldots, a_d)$, all points of the form $(y_1, \ldots, y_d)$ with $0 \leq abs(y_j) \leq abs(a_j)$, for all $j = 1, \ldots, d$ belong to the design D, where $abs(x)$ is the absolute value of x.*

Theorem 29 *Generalised echelon designs are minimal fan.*

Proof. An elegant proof exploits the notion of distraction and can be found in Robbiano and Rogantin (1998). □

For any integer $r \geq 1$ define the univariate polynomial

$$u_r(z) = \prod_{s=0}^{r-1} (z - s)$$

Then, for any multiple index $\alpha = (\alpha_1, \ldots, \alpha_d)$ define $g_\alpha(x) = \prod_{i=1}^{d} u_{\alpha_i}(x_i)$. Now a generalised echelon design can be defined in terms of a special set E of index vectors $\alpha^{(1)}, \ldots, \alpha^{(m)}$

$$E = \left\{\beta : 0 \leq \beta \leq \alpha^{(j)}, \text{ for some } j = 1, \ldots, m\right\}$$

Moreover, we may use unique $\alpha^{(j)}$ in the sense that for no $j' \neq j$, it is true that $\alpha^{(j')} \leq \alpha^{(j)}$. In this case, the following holds for any term-ordering τ

(i) the (reduced) Gröbner basis elements are $g_{\alpha^{(j)}}(x)$, for $j = 1, \ldots, m$

(ii) $\mathrm{LT}_\tau\left(g_{\alpha^{(j)}}(x)\right) = x^{\alpha^{(j)}}$, for $j = 1, \ldots, m$

(iii) $\mathrm{Est}_\tau(E) = \left\{x^\beta : \beta \in E\right\}$.

The only difference for non-equally spaced design with the same echelon structure is in (ii). If E' is such a design, we obtain

(iii') $\mathrm{Est}_\tau(E') = \left\{x^\beta : \beta \in E\right\}$.

For a generalised echelon design it is necessary to find an echelon design D with the same Est.

Example 49 [Continuation of Example 32] Echelon designs are a special case of generalised echelon design and thus are minimal fan. Moreover, an echelon design identifies only the order ideal whose pattern is the same pattern as the design. For the design in Example 32, the identifiable set is

$$\left\{1, \quad x_1, \quad x_1^2, \quad x_1^3, \quad x_2, \quad x_1 x_2, \quad x_1^2 x_2, \quad x_2^2\right\}$$

Definition 34 *Given d and N positive integers, the set of all order ideals in d indeterminates with exactly N terms is indicated by $\mathcal{G}(d, N)$.*

Definition 35 *A design is called* maximal (statistical) fan *if its fan is all $\mathcal{G}(d, N)$, where d is the number of factors in D and N the number of distinct points in D.*

Theorem 30 *A design, chosen randomly with respect to any absolutely continuous measure with respect to the Lebesgue measure, identifies all models in $\mathcal{G}(d, N)$ with probability one.*

Proof. Let D be a design. For all $E \in \mathcal{G}(d, N)$ the condition of identifiability (invertibility of the design matrix) $\det(Z(E, D)) = 0$ defines a variety in the $d \times N$ space of all coordinates of $D = \{x(i) : i = 1, \ldots, N\}$ which is of dimension less than $d \times N$. This follows from the linear independence of the monomials in any fan. Then, the set

$$\bigcup_{E \in \mathcal{G}(d,N)} \{D : \det(Z(E, D)) = 0\}$$

remains of dimension less than $d \times N$, since $\mathcal{G}(d, N)$ is finite. Any design D whose coordinates do not lie on this variety (technically, any point in the open set which is the union of the complement of the individual varieties

$\det(Z(E, D)) = 0)$ will have all $\det(Z(E, D)) \neq 0$. Since this holds with probability one, we are done. □

It is still a conjecture to prove that there are maximal fan designs supported on the integer grid. Nevertheless, a weaker but important result is presented in Theorem 31.

Example 50 This is an example of maximal fan design in any design size and number of factors. Let $\{q_1, \ldots, q_d\}$ be the first d prime numbers $\{1, 2, \ldots\}$. Then define $D = \{x(i)\}_{i=1}^{N}$ where

$$x_j(i) = q_i^{j-1} \quad (j = 1, \ldots, N)$$

Then consider the second row $(i = 2)$ of a typical design matrix, Z for an element of $\mathcal{G}(d, N)$ and the design D. The elements of this row are distinct because each entry represents a distinct prime power decomposition. Now all other rows of Z are distinct powers of this second row. This implies that Z is of Vandermonde type and therefore has non-zero determinant.

Theorem 31 *Given a positive integer N there exist designs that identify all models L where*

1. *L has N terms,*
2. *L is an order ideal,*
3. *there is a term-ordering with respect to which the terms in L are the smallest chain of length N.*

Proof. Consider a maximal fan design and apply the algorithm in Section 3.12. □

Example 51 Below are examples of maximal fan designs in two dimensions supported on the integer grid

N	D
3	$\{(0,0), (1,2), (2,1)\}$
4	$\{(0,1), (1,3), (2,0), (3,2)\}$
5	$\{(0,1), (1,3), (2,0), (3,4), (4,2)\}$
6	$\{(0,1), (1,5), (2,3), (3,0), (4,4), (5,2)\}$
7	$\{(0,1), (1,4), (2,2), (3,6), (4,0), (5,5), (6,3)\}$

3.11 Hilbert functions and fans for graded ordering

The Hilbert function, which plays an important role in the definition and computation of dimension for algebraic varieties, is only briefly introduced in Section 2.11. However, we can see that it leads to a pleasing result for the fan of an experimental design when the term-ordering is graded.

The definition of grading of a term-ordering is given in Definition 9. See also Definition 23.

Theorem 32 *For any design and any graded term-ordering τ the number of terms of $\mathrm{Est}_\tau(D)$ of a given order s is the same.*

Proof. Cox, Little and O'Shea (1997, Proposition 4, Section 9.3) states that for any graded ordering $\succ$ the Hilbert function of the monomial ideal I is the same. Thus simply specialize Item (i) of Theorem 18 to τ and remark that $\mathrm{Est}_\tau(D)$ consists precisely of monomials not in $\mathrm{LT}\,(\mathrm{Ideal}(D))$, that is $\mathbb{Q}[x_1, \ldots, x_d]/\mathrm{Ideal}(D)$. Then, for graded orderings the Hilbert functions are the same and so the number of terms of a given degree s, ${}^aHF_i(s) - {}^aHF_i(s-1)$, is the same for all graded orderings. □

This has a nice interpretation for the sub-fan of all leaves corresponding to graded orderings. We might call this the *graded fan*. Namely, every leaf has the same number of terms of a given degree.

In Section 2.3 we showed examples of gradings and, in particular, showed that some graded orderings are obtained by setting all the entries of the first row of the ordering matrix equal to one.

Example 52 We consider an example in $\mathbb{R}^3$. For our first ordering τ_1 we take tdeg with $x_1 \succ x_2 \succ x_3$ and the second ordering τ_2 defined by the matrix

$$\begin{bmatrix} 1 & 1 & 1 \\ -1 & 0 & -1 \\ -1 & -1 & 0 \end{bmatrix}$$

Take the design $\{(0,0,0),(1,1,1),(1,1,0),(2,0,2),(0,3,3)\}$. Calculations give

$$\mathrm{Est}_{\tau_1}(D) = \{1, x_1, x_2, x_3, x_3^2\} \text{ and } \mathrm{Est}_{\tau_2}(D) = \{1, x_1, x_2, x_3, x_1^2\}$$

and Theorem 32 is confirmed.

3.12 Subsets and algorithms

For a number of statistical reasons, we are interested in adding to or subtracting points from a design. The following result shows that elements of Est_τ are correspondingly added or subtracted.

Theorem 33 *Let $D_1 \subseteq D_2$ be two experimental designs then for the same term-ordering τ*

$$\mathrm{Est}_\tau(D_1) \subseteq \mathrm{Est}_\tau(D_2)$$

Proof. In what follows, $\mathrm{Ideal}(D)$ is the design ideal for the design D and $\{\mathrm{LT}_\tau(\mathrm{Ideal}(D))\}$ is the set of leading terms of $\mathrm{Ideal}(D)$ with respect to the term-ordering τ. The following relationships prove the theorem

$$\begin{aligned} D_1 \subseteq D_2 &\iff \mathrm{Ideal}(D_1) \supseteq \mathrm{Ideal}(D_2) \\ &\implies \{\mathrm{LT}_\tau(\mathrm{Ideal}(D_1))\} \supseteq \{\mathrm{LT}_\tau(\mathrm{Ideal}(D_2))\} \\ &\iff \mathrm{Est}_\tau(D_1) \subseteq \mathrm{Est}_\tau(D_2) \end{aligned}$$

Note that the last step uses the fact that for a design D, $\mathrm{Est}_\tau(D)$ is the complementary set of $\{\mathrm{LT}_\tau(\mathrm{Ideal}(D))\}$ equivalently of the set of leading terms of the Gröbner basis of $\mathrm{Ideal}(D)$. The second implication follows from the definition of $\{\mathrm{LT}_\tau(\mathrm{Ideal}(D))\}$. □

Corollary 6 *For any two designs D_1 and D_2 and a given term-ordering τ*
$$\mathrm{Est}_\tau(D_1 \cap D_2) \subseteq \mathrm{Est}_\tau(D_1) \cap \mathrm{Est}_\tau(D_2)$$
$$\subseteq \mathrm{Est}_\tau(D_1) \cup \mathrm{Est}_\tau(D_2) \subseteq \mathrm{Est}_\tau(D_1 \cup D_2)$$

Proof. The first and last inclusion follow from Theorem 33 by using the appropriate sets. Note that even if D_1 and D_2 are distinct, then the last inclusion is strict because both $\mathrm{Est}_\tau(D_1)$ and $\mathrm{Est}_\tau(D_1)$ contain 1 (at least). □

Theorem 33 implies that if we add points, one by one, to a design, so we add terms to Est. This can be turned into an algorithm for computing the successive terms of Est, which is statistical in flavor.

Let $D_N = \{a(1), \ldots, a(N)\}$ and $\mathrm{Est} = \{x^{\beta_1}, \ldots, x^{\beta_N}\}$ where we have considered the x^{β_k} to have been added sequentially. Also let
$$Z_\tau(D_N) = \left[(a(j))^{\beta_k}\right]_{j,k=1}^N$$
be the design matrix for the design D_N and the monomials
$$\{x^{\beta_k} : k = 1, \ldots, N\}$$
The next result considers the addition of a new point $a(N+1)$ to form
$$\mathrm{Est}_\tau(D_{N+1}) = \mathrm{Est}_\tau(D_N) \cup \{x^{\beta_{N+1}}\}$$

Theorem 34 *In adding a point $a(N+1)$ to a design D_N, the additional monomial $x^{\beta_{N+1}}$ such that $\mathrm{Est}(D_{N+1}) = x^{\beta_{N+1}} \cup \mathrm{Est}(D_N)$ satisfies*

1. $x^{\beta_{N+1}} \notin \mathrm{Est}(D_N)$
2. *The design matrix for $\mathrm{Est}_\tau(D_{N+1})$ and D_{N+1} is non-singular*
3. $x^{\beta'} \succ_\tau x^\beta$ *for any other β' satisfying Item 1 and Item 2 above.*

Proof. Item 2 must be satisfied for the set $\mathrm{Est}(D_{N+1})$ to be identifiable by D_{N+1}. Since $\mathrm{Est}_\tau(D_{N+1})$ must be an order ideal, we can see that the only candidate to be $x^{\beta_{N+1}}$ are $\{\mathrm{LT}(g_j)\}$ where $G = \{g_j\}$ is a Gröbner basis for D_N with respect to τ. Thus Item 1 is proved. We prove Items 2 and 3 by contradiction.

Let $x^{\beta_{N+1}}$ satisfy 1, 2 and 3 and x^γ another element of $\{\mathrm{LT}(g_j)\}$ and thus satisfying Items 1 and 2 but not Item 3. Let $\mathrm{Est}_\beta = \mathrm{Est}(D_N) \cup x^{\beta_{N+1}}$ and $\mathrm{Est}_\gamma = \mathrm{Est}(D_N) \cup x^\gamma$.

Now consider Item 2 and proceed by contradiction. Thus let β be defined as in the theorem so that $\gamma \neq \beta$, $x^\gamma \succ_\tau x^\beta$. Now x^β remains a leading term of some Gröbner basis element $g(x)$ of D_{N+1}, which we can write

$$g(x) = \theta_\beta x^\beta + \sum_{\alpha \in L \cup \gamma} \theta_\alpha x^\alpha$$

But then, since $x^\gamma \succ_\tau x^\beta$, we must have $\theta_\gamma = 0$. But since $g(x) = 0$ on D_{N+1} and $Est(D_N) \cup x^\beta$ is invertible over D_{N+1}, all the coefficients of $g(x)$ must be zero, which is a contradiction. □

Consider then adding $x^{\beta_{N+1}}$ terms to Est according Theorem 34. Let us define $g_N(x)$ to be the Gröbner basis element for Est whose leading term is $x^{\beta_{N+1}}$. Assume also that the data is given by

$$\{y_1, \ldots, y_N, y_{N+1}, \ldots\}$$

in the order corresponding to the addition of points. Let $p_N(x)$ be the interpolator of $\{a(N), y_N\}$. Then the following holds.

Corollary 7 *The following up-dating formula holds for the interpolators $\{p_N(x)\}$*

$$p_{N+1}(x) = p_N(x) + (y_{N+1} - p_N(a(N+1))) \frac{g_N(x)}{g_N(a(N+1))}$$

Proof. Since $g_N(x) = 0$ on D_N, $p_{N+1}(a(i)) = p_N(a(i)) = y_i$ $(i = 1, \ldots, N)$. But at $a(N+1)$, $p_{N+1}(a(N+1)) = y_{N+1}$ provided that $g_N(a(N+1)) \neq 0$. But the latter cannot happen because then $g_N(x) = 0$ on D_{N+1} and the fact that $Est_\tau(D_{N+1}) = x^\beta \cup Est(D_N)$ is non-singular on D_{N+1} would force $g_N(x) = 0$, similarly to the proof of Theorem 34. □

The algorithm is as follows

(i) $D_1 = a(1)$ and $Est_\tau(D_1) = \{1\}$

At the Nth step do

(ii) Add $a(N+1)$ to D_N to form D_{N+1}.

(iii) Update $Est_\tau(D_{N+1}) = Est_\tau(D_N) \cup x^{\beta_{N+1}}$ where $x^{\beta_{N+1}}$ is computed according to Theorem 34: the first "unused" $x^{\beta_{N+1}}$ for which the new design matrix is non-singular. Call the new matrix Z_{N+1}.

(iv) Construct $p_{N+1}(x)$ by $p_{N+1}(x) = [\theta_\alpha]^t [x^\alpha]$ where $[\theta_\alpha] = Z_{N+1}^{-1} Y_{(N)}$, where $Y_{(N)}$ is the vector $[y_1, \ldots, y_N]^t$ and $[x^\alpha]$ contains the elements of $Est_\tau(D_{N+1})$ in appropriate order.

(v) Construct (up to a scalar)

$$g_N(x) = p_{N+1}(x) - p_N(x)$$

This algorithm is very similar in structure to an algorithm in Abbott, Bigatti, Kreutzer and Robbiano (2000) and it is based on Gaussian elimination. It is convenient to discuss both algorithms via a tableau representation.

The basic tableau is in effect a developing version of the design matrix. We first represent an example and then give the steps more formally as above.

Let us consider the following design in $\mathbb{Q}^2$

$$D_4 = \{(0,0), (1,0), (0,1), (1,2)\}$$

and take the tdeg term-ordering with $x_2 \succ x_1$ as the initial order. In particular we have

$$\ldots \succ x_2^2 \succ x_1 x_2 \succ x_1^2 \succ x_2 \succ x_1 \succ 1$$

A starting tableau with points as columns and monomials as rows is given in Table 3.4

	(0,0)	(1,0)	(0,1)	(1,2)
1	1	1	1	1
x_1	0	1	0	1
x_2	0	0	1	2
x_1^2	0	1	0	1
$x_1 x_2$	0	0	0	2
x_2^2	0	0	1	4

(3.4)

This is the transpose of the usual design matrix.

As we proceed down the rows of the tableau the first three rows form a full rank matrix. However with the next row, x_1^2, the 4×4 matrix of the first four rows is singular. We can see this by a row reduction (subtract row 2 from row 4). The new tableau is

	(0,0)	(1,0)	(0,1)	(1,2)
1	1	1	1	1
x_1	0	1	0	1
x_2	0	0	1	2
$x_1^2 - x_1$	0	0	0	0
$x_1 x_2$	0	0	0	2
x_2^2	0	0	1	4

From the algorithm above we see that x_1^2 is not in $\text{Est}(D_4)$. However if, instead we consider $x_1 x_2$, then the sub-matrix formed by the four columns and rows 1, 2, 3 and 5 is non-singular and $x_1 x_2$ is our new element in Est.

Now consider the addition of a further design point $(2,1)$. Then, the tableau has five columns and becomes

	$(0,0)$	$(1,0)$	$(0,1)$	$(1,2)$	$(2,1)$
1	1	1	1	1	1
x_1	0	1	0	1	2
x_2	0	0	1	2	1
$x_1^2 - x_1$	0	0	0	0	3
$x_1 x_2$	0	0	0	2	2
x_2^2	0	0	1	4	1

The entry "3" in row 4 now indicates that indeed x_1^2 is a valid Est member. Only if we find a complete row of zeros (for all columns) would the term be prohibited. Suppose for example we has added the point $(1,3)$ instead of $(2,1)$, then the the element "3" would be replaced with "0". In that case (a row of zeros) we can write down the operations on the row $x_1^2 - x_1$ and claim that $g(x) = x_1^2 - x_1$ is an element of the Gröbner basis.

The algorithm is as follows. We start with a N-point design.

(i) Choose a term-ordering τ and line up the first N smallest monomials in the τ-ordering. Set up the tableau with a column for each point and for the moment an open-ended number of columns in the τ-ordering. Assume an N-point design. The first row is all 1's.

At the Nth step perform row reduction and

(ii) call each new row "good" if the sub-matrix formed by that row together with previous good rows is full rank.

(iii) Call the new row "bad" if the sub-matrix in (i) is not full rank, revealed by a row of zeros after row reduction.

(iv) Provided the number of good rows is less then N, each good row corresponds to a member of Est.

(v) Provided the stopping rule (see below) is not yet satisfied each bad row in (ii) gives a member of the Gröbner basis of the design ideal of the N points considered.

The stopping rule is based on the relationship between Est and the Gröbner basis: stop if a new bad row corresponds to a monomial divisible by the leading term of the Gröbner basis generated by any previous bad row. All previous bad rows yield the full Gröbner basis. Another stopping rule is when the number of points the design and the number of terms in Est are the same.

We can see from the example that a suitable sequential reordering of points (columns) is guaranteed to produce an ordering of points such that the sequential algorithm would not need to backtrack to revisit a previous "failed" monomial.

3.13 Regression analysis

The formal link with regression in statistics should by now be clear. For a design D and a regression model

$$Y = Z\theta + \varepsilon$$

the design matrix $Z = [x^\alpha]_{x \in D; \alpha \in L}$, $\mathrm{Est}_\tau(D) = \{x^\alpha : \alpha \in L\}$ is non-singular and provides a saturated model. We collect together a series of notes on the use of the methods in practical regression. This is based on the author's joint work, particularly in industrial settings. It should be combined with and draw on the special constructions and algorithm of the last section. For some expansion of the issues discussed here, see Giglio, Riccomagno and Wynn (2000).

3.13.1 Submodels

Any sub-model of $\mathrm{Est}_\tau(D)$ that is $M = \{x^\alpha : \alpha \in L' \subset L\}$ is also identifiable with the design D and can be used (as can Est) as input to stepwise regression. All the standard theory of the linear model is applicable.

3.13.2 Orthogonal polynomials

A method used extensively by the authors is to compute orthogonal polynomials with respect to the term-ordering. Thus define

$$Y = W\phi + \varepsilon$$

where $W = Z_\tau U_\tau^{-1}$ and U_τ is the unique upper triangular (Cholesky) matrix such that $Z_\tau^t Z_\tau = U_\tau^t U_\tau$. Then

$$[h_\alpha(x)]_{\alpha \in L'} = \left[U_\tau^t\right]^{-1} [x^\alpha]_{\alpha \in L'}$$

are orthogonal polynomials with respect to D. It is natural to order the columns of X_τ in τ-ordering for then

$$h_\alpha(x) = \sum_{\alpha \succ_\tau \beta} c_{\alpha, \beta} x^\beta$$

and the model is written

$$\sum_{\alpha \in L'} \phi_\alpha h_\alpha(x)$$

Then (under the usual regression assumptions) the minimum variance linear unbiased estimates of the ϕ_α are $[\hat\phi_\alpha] = h[\hat\theta_\alpha]$ where $\hat\theta_\alpha$ are the least-squares estimates. The contribution to the usual regression sum of squares of each $\hat\phi_\alpha$ is simply $\hat\phi_\alpha^2$. Orthogonality means that the $\hat\phi_\alpha$ are independent under standard assumptions and can be used for standard half-normal plots and similar analysis.

3.13.3 Augmentation of designs

Theorem 34 is useful in augmentation of design with new points as may occur for example in optimal experimental design or sequential experimental design more generally. The method contains a guarantee that all existing models generated by a given τ will remain estimable by the augmented design. Note that it is not true in general that the orthogonal polynomials of the previous section remain the same.

3.13.4 Favored orderings and blocked orderings

Put simply, when a particular factor is small in the initial ordering it tends to be favored in the allocation of monomial terms in $\text{Est}_\tau(D)$. The lex ordering is a radical version of this feature. It is common in analysis to perform two- or multi-step procedures which would first screen the factors using a simple analysis such as testing only for linear terms and then reward the significant factors by putting them higher in the term-ordering, perhaps leaving out non-significant factors altogether. The combination of such strategies with the ability to construct quite general orderings and using the "blocking" of the last section seems to be powerful.

In general one can choose the whole structure of the term-ordering in order to force the Est set to include some terms, if the presence of the terms is compatible with the design.

3.14 Non-polynomial models

Other types of models can be put into the algebraic setting of this book. The important necessary condition is that the considered set of models forms a ring. In particular, a one-to-one correspondence between the class of models and a quotient space of a polynomial ring is required. This is made clear by considering the Fourier regression case. See Caboara and Riccomagno (1998).

Definition 36 *Let A be a finite subset of the d-dimensional integer vectors. The* Fourier *(or trigonometric) regression model based on A is defined as the linear model*

$$Y(x) = \sum_{h \in A} a_h \exp\left(2\pi i h^t x\right) + \varepsilon(x)$$

where a_h is in $k(i)$, the usual field of constants extended with the imaginary unit, i.

The set A plays the role of the frequencies. In general the models in Definition 36 contain a non-zero real part and a non-zero imaginary part. The following definition describes the set of Fourier models with a zero imaginary part.

NON-POLYNOMIAL MODELS

Definition 37 *Let A^+ be a finite set of d-dimensional integer vectors not containing zero and such that $h \in A^+$ implies $-h \notin A^+$. A complete Fourier (or trigonometric) model is defined as*

$$Y(x) = \theta_0 + \sqrt{2} \sum_{h \in A^+} \left[\theta_h \sin(2\pi h^t x) + \phi_h \cos(2\pi h^t x) \right]$$

$$= \alpha_0 + \sum_{h \in A^+} \left[\beta_h \exp(2\pi i h^t x) + \delta_h \exp(2\pi i h^t x) \right]$$

where

$$\alpha_0 = \theta_0$$

$$\beta_h = \frac{\phi_h - i\theta_h}{\sqrt{2}}$$

$$\delta_h = \frac{\phi_h + i\theta_h}{\sqrt{2}}$$

Heuristically the one-to-one correspondence mentioned above is realized by using the identities $s_i = \sin(2\pi x_i)$ and $c_i = \cos(2\pi x_i)$ for all $i = 1, \ldots, d$. This idea is formalised in the following theorem.

Theorem 35 *There is a one-to-one correspondence between the set of all Fourier models and the polynomial quotient ring*

$$k(i)[c_1, \ldots, c_d, s_1, \ldots, s_d] / \langle c_1^2 + s_1^2 - 1, \ldots, c_d^2 + s_d^2 - 1 \rangle$$

Proof. See Caboara and Riccomagno (1998). $\square$

The commonly used computer algebra packages do not yet include a version of the Buchberger algorithm for complex numbers. An ad hoc procedure for Fourier models has been implemented in CoCoA.

Another example is given by regression models based on increasing exponentials, for example

$$Y(x) = \theta_0 + \theta_1 \exp(x_1) + \theta_2 \exp(x_2) + \theta_3 \exp(x_1 + x_2) + \varepsilon$$

where ε is a random error with zero mean.

Definition 38 *The set of exponential regression models is defined as*

$$\mathcal{E} = \left\{ \sum_{h \in A} \theta_h \exp(h^t x) : A \subset \mathbb{Z}_+^d, A \text{ is finite and } \theta_h \in k \right\}$$

where k is the usual coefficient field.

Theorem 36 *The set $\mathcal{E}$ is in one-to-one correspondence with the polynomial ring $k[t_1, \ldots, t_d]$.*

Proof. The proof of the elementary fact that the relationship $t_i = \exp(x_i)$ for $i = 1, \ldots, d$ defines a ring isomorphism between $\mathcal{E}$ and $k[t_1, \ldots, t_d]$ is left to the reader. $\square$

CHAPTER 4
Two-level factors: logic, reliability, design

4.1 The binary case: Boolean representations

There are close connections between two-level factorial designs and various algebraic structures which employ binary representations. We will try to arrive at a point where moving between the representations will give additional insight.

The starting point is a set of variables $x_1, \ldots, x_d$ which are binary. We code these two levels as $\{0, 1\}$. Following the analysis of the previous chapters, we represent the set of values of a vector $x = (x_1, \ldots, x_d)$ as a 2^d full factorial design, $D_{2^d} = \{0, 1\}^d$. This can be represented as the solutions of

$$\{x_i(x_i - 1) = 0 : i = 1, \ldots, d\}$$

Next, we specialize the theory of Chapter 3 to the binary case.

Definition 39 *Given a design, a fraction is simply a proper subset of the design. In particular, a fraction D_N is a subset of D_{2^d} of size N. Sometimes we simply write D.*

According to the Gröbner basis theory, given a term-ordering τ, the fraction D has a Gröbner basis representation: if the reduced Gröbner basis of Ideal(D) with respect to τ is

$$\{g_j(x) : j = 1, \ldots, t\}$$

then D is given by the solutions of

$$\{g_j(x) = 0 : j = 1, \ldots, t\}$$

Corresponding to D and given a term-ordering τ, there is a set of estimable terms

$$\text{Est}_\tau(D) = \{x^\alpha : \alpha \in L\} \tag{4.1}$$

which forms a vector space basis for $k[x_1, \ldots, x_d]/\text{Ideal}(D)$, the quotient vector space, where Ideal(D) is the design ideal: $\langle g_j(x) : j = 1, \ldots, t\rangle$.

The special feature of the binary case is that the terms of $\text{Est}_\tau(D)$ are multi-linear, that is square free. Thus the set L in (4.1) consists of binary vectors. We recall also that $\text{Est}_\tau(D)$ (or equivalently L) is an order ideal.

Thus, for example, if $x_1 x_2 x_3$ is in L so are

$$1, \quad x_1, \quad x_2, \quad x_3, \quad x_1 x_2, \quad x_2 x_3, \quad x_1 x_3$$

Note also that when $D = D_{2^d}$, D and L coincide because D_{2^d} is echelon and thus minimal fan (see Section 3.10). Indeed the reduced Gröbner basis for D_{2^d} is $\{x_i^2 - x_i : i = 1, \ldots, d\}$, for all term-orderings.

Definition 40 *A Boolean algebra* $\mathcal{B}(\vee, \wedge, ^-, 0, 1)$ *is a set* $\mathcal{B}$ *with two binary operations* $\vee$ *and* $\wedge$, *a complement* $^-$, *a null element* 0 *and a unit* 1. *The set satisfies the commutative and associative properties over* $\vee$ *and* $\wedge$, *the two distributive properties*

$$a \wedge (b \vee c) = (a \wedge b) \vee (a \wedge c) \quad a \vee (b \wedge c) = (a \vee b) \wedge (a \vee c)$$

and

$$a \vee 0 = 0 \quad a \wedge 1 = a \quad a \vee \bar{a} = 1 \quad a \wedge \bar{a} = 0$$

$(a, b, c \in \mathcal{B})$.

Example 53 We can relate this directly to the design theory using a standard set representation, as shown below

$$
\begin{array}{rcl}
1 & \leftrightarrow & D_{2^d} \\
\text{element } a & \leftrightarrow & \text{design } D \\
0 & \leftrightarrow & \text{empty set } \emptyset \\
\vee & \leftrightarrow & \cup \\
\wedge & \leftrightarrow & \cap \\
\bar{a} & \leftrightarrow & D_{2^d} \setminus D \text{ (written } \bar{D})
\end{array}
$$

With this correspondence designs, as subsets of a full factorial design, become a Boolean algebra.

In Chapters 2 and 3 we showed how to construct a polynomial interpolator over a set by using $\text{Est}_\tau(D)$. Thus any function, g over D_{2^d} can be expressed uniquely as a polynomial

$$g = \sum_{\alpha \in L_{2^d}} \theta_\alpha x^\alpha$$

where $L_{2^d} = \{0, 1\}^d$.

A further representation of the design is with indicator functions.

Definition 41 *The* indicator (polynomial) function *of a binary design* $D \subseteq D_{2^d}$ *is*

$$f_D(x) = \sum_{\alpha \in L_{2^d}} \theta_\alpha x^\alpha$$

$$\text{where } f_D(x) = \begin{cases} 1 & \text{if } x \in D \\ 0 & \text{if } x \in \bar{D} \end{cases}$$

THE BINARY CASE: BOOLEAN REPRESENTATIONS

A formal way of expressing this is as a member of the quotient ring $k[x_1,\ldots,x_d]/\text{Ideal}(D_{2^d})$ with the additional restriction $f_D^2 = f_D$.

Example 54 The indicator function of the fraction $D = \{(0,1),(1,0)\}$ of D_{2^2} is $f_D(x_1, x_2) = x_1 + x_2 - 2x_1x_2$ in $\mathbb{Q}[x_1, x_2]/\langle x_1^2 - x_1, x_2^2 - x_2\rangle$. Over the coefficient field $\mathbb{Z}_2$ the indicator function of D is $x_1 + x_2$.

To obtain the Boolean algebra representation we simply note that for two designs D_1 and D_2 and corresponding indicator functions $f_{D_1}(x)$ and $f_{D_2}(x)$, we have

$$f_{D_1 \cup D_2}(x) = \max\left(f_{D_1}(x), f_{D_2}(x)\right) = f_{D_1}(x) + f_{D_2}(x) - f_{D_1}(x)f_{D_2}(x)$$
$$f_{D_1 \cap D_2}(x) = f_{D_1}(x)f_{D_2}(x)$$
$$f_{\bar{D}_1}(x) = 1 - f_{D_1}(x)$$
$$f_{D_{2^d}}(x) = 1$$
$$f_\emptyset(x) = 0$$

for $x \in D_{2^d}$ and where $\emptyset$ is the empty set.

Given a design D we can construct its polynomial indicator function f_D as follows. First write down the indicator for a single point $\omega \in D$

$$f_\omega(x) = \prod_{i:\omega_i=1} x_i \prod_{i:\omega_i=0} (1-x_i)$$

then clearly

$$f_D(x) = \sum_{\omega \in D} f_\omega(x) = \sum_{\omega \in D} \prod_{i:\omega_i=1} x_i \prod_{i:\omega_i=0} (1-x_i)$$

Note that $f_D(x)$ contains, typically, square-free (multi-linear) monomials up to the highest order: $\prod_{i=1}^d x_i$. This representation is unique by construction. In general the complexity of the monomial terms depends on the complexity of the design D, a point we shall return to later.

We can thus move backwards and forwards between the set representation and the polynomial indicator function representation for designs.

We give two more alternative representations of a design, as a logical proposition and as a binary string relating to the vertices of the unit hypercube. A design D (or set) can be further interpreted as a logical proposition by the translation

$$\cup \leftrightarrow \vee \text{ (or)}$$
$$\cap \leftrightarrow \wedge \text{ (and)}$$
$$- \leftrightarrow \neg \text{ (not)}$$

Next, one can also represent D as a set of binary strings of length d (the coordinate points). Then, a complex Boolean expression using indicator

functions such as
$$f = \max(f_1 f_2, f_3) = f_1 f_2 + f_3 - f_1 f_2 f_3$$
$$= (f_1 + f_3 - f_1 f_3)(f_2 + f_3 - f_2 f_3) = \max(f_1, f_3) \max(f_2, f_3)$$
represents the corresponding evaluation of the truth of the underlying proposition
$$D = (D_1 \wedge D_2) \vee D_3 = (D_1 \vee D_3) \wedge (D_2 \vee D_3)$$

Finally, order the corners of the 2^d-full factorial design D_{2^d}. A design can be represented by a binary string of length 2^d. The idea is made clear with an example. For $d = 2$ fix an anti-clockwise order on the points of the $\{0,1\}^2$ grid starting from $(0,0)$, that is $(0,0)$, $(1,0)$, $(1,1)$ and $(0,1)$ last. Below the representation of three designs is given

$$D_1 = \{(0,0),(1,0)\} \qquad = (1,1,0,0)$$
$$D_2 = \{(0,0),(1,1),(0,1)\} = (1,0,1,1)$$
$$D_3 = \{(1,1),(0,1)\} \qquad = (0,0,1,1)$$

The design $D = (D_1 \cap D_2) \cup D_3$, above corresponds to $(1,0,1,1)$. Putting D_1, D_2, D_3 and D together gives a truth table where T and F can be interpreted as the truth of the proposition

	$(0,0)$	$(1,0)$	$(0,1)$	$(1,1)$
D_1	T	T	F	F
D_2	T	F	T	T
D_3	F	F	T	T
D	T	F	T	T

4.2 Gröbner bases and Boolean ideals

Given a term-ordering τ and a coefficient field including two different elements like 0 and 1, a design D is represented by a unique reduced Gröbner basis, G, with respect to the τ term-ordering. Write the Gröbner basis, G and the corresponding ideal as

$$\text{Ideal}(D) = \langle g(x) : g \in G \rangle$$

Let us repeat from the previous section that typically amongst the leading terms of G there are monomials of the type x_i^2 and in Ideal(D) there are the polynomials $x_i^2 - x_i$ for $i = 1, \ldots, d$. However if D is a proper subset of D_{2^d} then for all $g \in G$ the leading term of g is either x_i^2 for a certain $i = 1, \ldots, d$ or a multi-linear term.

Example 55 Let $d = 3$ and consider the set

$$D = \{(1,1,0),(1,0,1),(0,0,1),(1,1,1)\}$$

The Gröbner basis with respect to the tdeg term-ordering is given by the following polynomials

$$\begin{cases} x_3^2 - x_3 \\ x_2 x_3 - x_2 - x_3 + 1 \\ x_1 x_3 - x_1 - x_3 + 1 \\ x_2^2 - x_2 \\ x_1 x_2 - x_2 \\ x_1^2 - x_1 \end{cases}$$

with leading terms

$$\{x_3^2, \quad x_2 x_3, \quad x_1 x_3, \quad x_2^2, \quad x_1 x_2, \quad x_1^2\}$$

Now suppose that we can assume that $D \subseteq D_{2^d}$, that is, we know in advance that the design is binary. Then, restricting to D_{2^d} we only need use the multi-linear leading terms of the Gröbner basis and state that D is given as the solution in D_{2^d} of

$$\begin{cases} x_2 x_3 - x_2 - x_3 + 1 = 0 \\ x_1 x_3 - x_1 - x_3 + 1 = 0 \\ x_1 x_2 - x_2 = 0 \end{cases}$$

However, we now have a competitor for the expression of D to the alternative expression obtained by simply putting its polynomial indicator function equal to unity

$$1 - f_D(x) = 0$$

In the example this is

$$1 - (x_1 x_2 - x_2 x_3 + x_3) = 0 \tag{4.2}$$

The alternative forms are representations of Boolean ideals and corresponding Boolean varieties. By this we mean the set of all indicator functions which are zero on D and unity on $D_{2^d} \setminus D = \bar{D}$.

Let us be a little more precise and link the discussion more closely to the Boolean algebra formulation, see Halmos and Givant (1998).

Definition 42 *An ideal $\mathcal{I}$ in a Boolean algebra $\mathcal{B}$ is a subset of $\mathcal{B}$ with the properties*

1. $0 \in \mathcal{I}$

2. if $f \in \mathcal{I}$ and $g \in \mathcal{I}$ then $f \vee g \in \mathcal{I}$

3. if $f \in \mathcal{I}$ and $g \in \mathcal{B}$ then $f \wedge g \in \mathcal{I}$.

Now consider $D \subseteq D_{2^d}$. Let D_1 and D_2 be any two sets containing D then $D \subseteq D_1 \cap D_2$. If E is any other set in D_{2^d} then $D \subseteq D_1 \cup E$. Now let $g_D = 1 - f_D$ where f_D is the indicator function of D and similarly for D_1, D_2 and E. Then, $\max(g_{D_1}, g_{D_2}) = 0$ on D and $g_{D_1} g_E = 0$ on D. Thus, with zero identified as the zero indicator we recapture either in the set theory or in the polynomial representation the notion of a Boolean ideal.

This is sometimes described by saying that D is the *kernel* of the Boolean homomorphism
$$g_D = 1 - f_D$$
There is an even simpler interpretation of the ideal using an inequality: $D \subseteq D_1$ means that g_{D_1} is in the ideal generated by G_D. But it is enough to write
$$f_{D_1} \geq f_D$$
for the two indicator functions. Thus we can, as is well known, work equivalently within the theory of partial ordering and lattices.

4.3 Logic and learning

Recall that a logical proposition can be considered as a subset D of D_{2^d}, the set of all elementary propositions (also called outcomes, atoms, etc). To repeat, D can be considered as an element of a Boolean algebra or a Boolean function $f_D(\omega)$, uniquely expressible as a polynomial consisting of square-free monomials.

We may fix a particular D and consider it to be the set of all true elementary propositions. In many situations in logic we may wish to determine D from other, partial information. In particular, in certain models of learning we may be presented with training or test examples which state the truth or falsehood of certain test cases. These can be presented as a set of N pairs
$$(a(i), y_i) \quad (i = 1, \ldots, N)$$
where y_i is the truth value of $a(i) \in D_{2^d}$. Since we may interpret the truth of a statement $a(i)$ as the statement that $a(i) \in D$ we can evaluate y_i by
$$y_i = f_D(a(i)) \quad (i = 1, \ldots, N)$$
where $f_D(a(i))$ (as usual) is the polynomial indicator for D, giving $y_i = 0$ or 1.

In this case, we may call a (current) *hypothesis* any indicator function $f_{D'}$ which is consistent with the current truth, namely the set of data so far collected: $(a(i), y_i)$ $(i = 1, \ldots, N)$. Such hypothesis satisfies
$$y_i \equiv f_{D'}(a(i)) = f_D(a(i)) \quad (i = 1, \ldots, N)$$
Define
$$D_N = \{a(i) : a(i) \in D, i = 1, \ldots, N\}$$
$$\bar{D}_N = \{a(i) \in D_{2^d} : a(i) \notin D\}$$
Then, the above statement can be written as an approximation for the hypothesis, D'
$$D_N \subseteq D' \quad \text{and} \quad \bar{D}' \subseteq \bar{D}_N \tag{4.3}$$

One can see from the discussion in the last section that (4.3) can be interpreted as a statement about inclusion of the corresponding ideals, either the set version or the indicator function version.

See Anthony and Biggs (1997) for an introduction to computational learning.

4.4 Reliability: coherent systems as minimal fan designs

The development in Section 4.1 is a useful platform for doing elementary probability. We shall give a fuller development in Chapter 5. For the moment we simply consider a design D as an event in the probability space of all events ($D_{2^d} = \Omega$ in probability notation). Then we simply obtain the Boolean ring of events. All that remains is to attach probabilities $\{p(x)\}$ so that

$$0 \leq p(x) \leq 1 \text{ and } \sum_{x \in D_{2^d}} p(x) = 1$$

In Chapter 5 we shall extend the theory to give algebraic representations for the $p(x)$ themselves. We give now a fresh application of the ideas to reliability theory, see Barlow (1998).

Definition 43

1. *We call* system *a set S which consists of d components each of which can fail or not fail coded $1, 0$, respectively.*
2. *An outcome is a binary string of length d describing the failure or otherwise of each component.*
3. *A* failure outcome *is an outcome which leads to the failure of the whole system S.*
4. *The* failure event *(perhaps we should say the* maximal failure event*) is the set of all failure outcomes. We shall denote it by D because it has (again) the structure of a design: a subset of D_{2^d}, the space of all outcomes.*

In simple reliability the (often criticized) assumption that each component fails independently is sometimes made. More generally we consider a (joint) probability distribution given by $\{p(x) : x \in D\}$. In either case, the probability of the system failure is

$$p(D) = \sum_{x \in D} p(x)$$

A coherent system is one for which changing a non-failed component to failed for any failure outcomes still yields a failure outcome. Thus, for example, for a four component system if $(1, 1, 0, 0)$ is a failure outcome, so are $(1, 1, 1, 0), (1, 1, 0, 1)$ and $(1, 1, 1, 1)$.

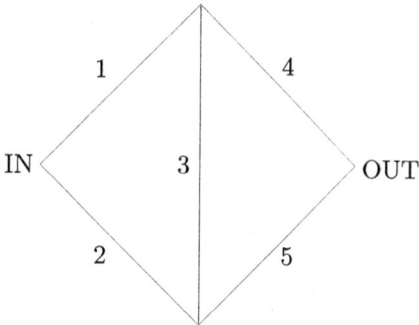

Figure 4.1 *An input/output system.*

Definition 44 *Let D_{2^d} represent a system with d components and let $D \subseteq D_{2^d}$ be a failure event. The system is said to be* coherent *if $x \in D$ implies $x' \in D$ for any $x' \geq x$ (in the usual entry-wise ordering: $x'_i \geq x_i$, $i = 1, \ldots, d$).*

The classical definition of coherence also requires that every component is non-redundant in the sense that the failure event contains the failure of any component at least once.

There is a useful connection between coherent systems and the minimal fan design in Section 3.10. The easiest way to see this is to (temporarily) reverse the coding to $0 \leftrightarrow 1$. Then, provided, there is at least one non-empty failure outcome we obtain exactly an echelon design as introduced in Example 32.

Example 56 Figure 4.1 shows a reliability network in standard form with edges as components and clearly IN and OUT stand for input and output. The system fails if the edges are cut so that there is no path from IN to OUT. Table 4.1 shows the cuts that give the failure event of the system. In the $\{0, 1\} \equiv \{\text{not fail, fail}\}$ coding the failure outcomes are given in Table 4.1. Reversing 0 and 1 yields an echelon design. Note that the minimal fan property is unaltered by the coding.

Definition 45 *For a coherent system a* minimal failure outcome *x is one for which any $x' \neq x$ for which $x' \leq x$ is not a failure outcome.*

Example 57 For the above network example the minimal failure outcomes are 11000, 00011, 10101 and 01110; these are coded as cuts 12, 45, 135 and 234 respectively.

Definition 46 *For a coherent system a* minimal non-failure outcome *x is a non-failure outcome for which any $x' \neq x$ with $x' \geq x$ is a failure outcome.*

Table 4.1 Cuts and failure event for the system in Figure 4.1.

12	45	135	11000	00011	10101
123	145	234	11100	10011	01110
124	245		11010	01011	
125	345		11001	00111	
1234	1345		11110	10111	
1235	2345		11101	01111	
1245			11011		
12345			11111		

Example 58 For the example in Figure 4.1 the minimal non-failure outcomes are 01101, 10110, 01010, 10001; these are coded as paths 14, 25, 135, 234 respectively. A *minimal path*, here, means a sequence of connected edges from input to output such that if an edge is out (fails) the path is broken and the system fails.

The set $\bar{D} = D_{2^d} \setminus D$ (where D is the failure event) is the non-failure event of which the minimal non-failure outcomes are members.

The connection with minimal fan design yields benefits. Considering D and $\bar{D}$ for a coherent system as designs we can construct their Est sets. By the minimal fan property they are independent of the term-ordering τ chosen. Thus for any τ

$$\mathrm{Est}_\tau(D) = \{x^\alpha : \alpha \in L\}$$

where $L = \{\alpha : \alpha_i = 1 - a_i, \text{ for } i = 1, \ldots, d \text{ and } a \in D\}$ and

$$\mathrm{Est}_\tau(\bar{D}) = \{x^\alpha : \alpha \in \bar{D}\}$$

We see that $\bar{D}$ is an echelon design and thus minimal fan in the standard orientation. Therefore it provides the Est term exponents for itself. Any function on D or $\bar{D}$, such as a cost dependent on failure, can be written as

$$D: \quad f = \sum_{\alpha \in L} \theta_\alpha x^\alpha$$

$$\bar{D}: \quad g = \sum_{\alpha \in \bar{D}} \phi_\alpha x^\alpha$$

4.5 Inclusion-exclusion and tube theory (with D. Naiman and B. Giglio)

Throughout this section we consider coherent systems so that the failure events D and non-failure events $\bar{D}$ are minimal fan.

Let us return to the use of indicator polynomials. Let
$$f_D(x) = \sum_{\alpha \in D_{2^d}} \theta_\alpha x^\alpha$$
be the indicator function for the failure event so that
$$f_D(x) = \begin{cases} 1 & x \in D \\ 0 & x \in \bar{D} \end{cases}$$
and recall that it is essentially a unique polynomial interpolator over D_{2^d}. The standard inclusion-exclusion identity for failure is based on the minimal failure outcomes, according to Definition 45. Let us call the collection of minimal failure outcomes F. Then for each $\omega \in F$ define the "quadrant"
$$Q_\omega = \{\omega' : \omega' \geq \omega, \quad \omega \in D_{2^d}\}$$
It is clear that
$$D = \bigcup_{\omega \in F} Q_\omega$$
Since the system is coherent for any $\omega \in F$, $Q_\omega \subseteq D$ the failure event. Thus Q_ω defines the set of all failure outcomes which include all failed components indicated by ω. Let F_r be the set of all subsets of F of size r and note that $F_1 = F$. The following is obtained simply by applying the usual inclusion-exclusion lemma. Note that the polynomial indicator of Q_ω is simply
$$q_\omega(x) = x^\omega = x_1^{\omega_1} \ldots x_d^{\omega_d}$$

Theorem 37 *For a coherent system with failure event D and minimal failure outcomes $F \subseteq D$, let $q_\omega(x)$ be the indicator function of $Q_\omega = \{\omega' \in D_{2^d} : \omega' \geq \omega\}$. Then, the indicator for the failure event is*
$$f_D(x) = \sum_{\omega \in F = F_1} q_\omega(x) - \sum_{(\omega^{(1)}, \omega^{(2)}) \in F_2} q_{\omega^{(1)}}(x) q_{\omega^{(2)}}(x) + \ldots (-1)^{\#F-1} \prod_{\omega \in F} q_\omega(x)$$
where $\#F$ is the cardinality of F.

There have been several attempts to obtain simplified versions of this formula. The importance stems from the following: each $q_\omega(x)$ or equivalently each Q_ω covers all events $\omega' \geq \omega$. Thus let $\omega = (\omega_1, \ldots, \omega_d)$ be a minimal failure event, let $\{p(\omega)\}$ be the failure distribution and let X be the corresponding variable in D_{2^d}. Then
$$\text{Prob}\{Q_\omega\} = \mathrm{E}(q_\omega(X)) = \mathrm{E}(X^\omega)$$

Similarly
$$\text{Prob}\left(Q_{\omega^{(1)}} \bigcap Q_{\omega^{(2)}}\right) = \text{E}\left(q_{\omega^{(1)}} q_{\omega^{(2)}}\right)$$
$$= \text{E}\left(X^{\omega^{(1)} \vee \omega^{(2)}}\right)$$

where $\omega^{(1)} \vee \omega^{(2)} = \left(\max(\omega_1^{(1)}, \omega_1^{(2)}), \ldots, \max(\omega_d^{(1)}, \omega_d^{(2)})\right)$ and so on for higher-order intersections.

Given a failure distribution $\{p(\omega)\}$ on D_{2^d}, the probability of failure is given by taking expectation of the inclusion-exclusion formula of Theorem 37 (with x replaced by X)

$$\text{P}(D) = \sum_{\omega \in F_1} \text{P}(Q_\omega) - \sum_{(\omega^{(1)}, \omega^{(2)}) \in F_2} \text{P}\left(Q_{\omega^{(1)}} \bigcap Q_{\omega^{(2)}}\right)$$
$$+ \ldots (-1)^{\#F-1} \text{Prob}\left(\bigcap_{\omega \in F} Q_\omega\right)$$

This is the usual formula for probability of failure based on "cuts" in system reliability.

Associated with the inclusion-exclusion formulae are inequalities. We state these for the indicator functions. Thus, for s odd ($s = 1, \ldots, \#F$)

$$f_D(x) \leq \sum_{r=1}^{s} (-1)^{r-1} \sum_{(\omega^{(1)}, \ldots, \omega^{(r)}) \in F_r} q_\omega(1) \cdots q_\omega(r)$$

and with reversed inequalities when s is even. Again, by taking expectation we obtain the probability version.

Coherent systems have the key property that the set of Q_ω which appear in the basic inclusion-exclusion lemma is very much restricted. Since each such Q_ω has a monomial indicator $q_\omega(x) = x^\omega = x_1^{\omega_1} \ldots x_d^{\omega_d}$ there is considerable reduction in complexity in inclusion-exclusion identities. Moreover, the standard inclusion-exclusion inequalities described above can be replaced by sharper inequalities with reduced complexity. Recent results are based on discrete tube theory which we now develop briefly. This is based on Giglio, Riccomagno and Wynn (2000) and Naiman and Wynn (2000). See also recent work by Dohmen (1999).

For the above example in binary notation the minimal failure outcomes are 11000, 00011, 10101 and 01110. The indicator function for the failure event given in Table 4.1 is easily computed as

$$f = x_1 x_2 + x_4 x_5 + x_1 x_3 x_5 + x_2 x_3 x_4$$
$$- x_1 x_2 x_4 x_5 - x_1 x_2 x_3 x_5 - x_1 x_2 x_3 x_4 - x_1 x_3 x_4 x_5$$
$$- x_2 x_3 x_4 x_5 + 2 x_1 x_2 x_3 x_4 x_5$$

Note that it uses only eleven monomials (counting multiplicities) out of a maximum of $2^5 = 32$ (including 1).

Definition 47 *An abstract simplicial complex is a collection $\mathcal{S}$ of non-empty subsets of an index set $\mathcal{I}$ such that for any $I \in \mathcal{S}$ and $J \subset I$ with $J \neq \emptyset$ we have $J \in \mathcal{S}$. The dimension of the simplicial complex is the maximal cardinality of I.*

The r-tuples of $\mathcal{S}$, that is the sets of size r, are called $r-1$ simplices. Thus a simplex of $\mathcal{S}$ has the property that all its sub-simplices also lie in $\mathcal{S}$.

The basic idea of tube theory is that, in some circumstances, we (i) index the sets we are interested in a special way (in our case, these are the quadrants $Q_\omega \in F$) and then (ii) identify a simplicial complex satisfying a special "discrete tube" property and finally (iii) claim inclusion-exclusion identities and inequalities only using terms associated with the simplicial complex.

Definition 48 *An abstract tube is a collection of subsets $\mathcal{A} = \{A_i : i \in \mathcal{I}\}$ of a set $\mathcal{X}$ where the index set $\mathcal{I}$ has a simplicial complex $\mathcal{S}$ such that for any $x \in \mathcal{X}$ the sub-simplicial complex*

$$\mathcal{S}_\omega = \left\{ I \in \mathcal{S} : x \in \bigcap_{i \in I} A_i \right\}$$

is contractible (as an abstract simplicial complex).

Here contractible means having the same "homotopy type" as a point, that is, roughly stated, it has a geometric realisation which can be continuously shrunk to a point.

The next result gives general inclusion-exclusion identities and inequalities for discrete tubes.

Theorem 38 *Let $\{\mathcal{A}, \mathcal{I}, \mathcal{S}\}$ be an abstract discrete tube and let $\mathcal{I} = n$. If $\chi(A)$ denotes the indicator function of a set A, then*

$$\chi(\cup_{i \in I} A_i) = \sum_{r=1}^{m} (-1)^{r-1} \sum_{\substack{\#I=r \\ I \in \mathcal{S}}} \chi(\bigcap_{i \in I} A_i)$$

$$\leq \sum_{r=1}^{s} (-1)^{r-1} \sum_{\substack{\#I=r \\ I \in \mathcal{S}}} \chi(\bigcap_{i \in I} A_i) \ (s \ odd)$$

$$\geq \sum_{r=1}^{s} (-1)^{r-1} \sum_{\substack{\#I=r \\ I \in \mathcal{S}}} \chi(\bigcap_{i \in I} A_i) \ (s \ even)$$

$(1 \leq s < m)$.

The maximal value m of $\#I$, $I \in \mathcal{S}$ is called the *depth* of the tube and is the dimension of the simplicial complex (plus one).

First we index the set of minimal failure points F in some way

$$F = \left\{\omega^{(1)}, \ldots, \omega^{(n)}\right\}$$

To construct a discrete tube, we need a mean of deciding which intersections of $\{Q_\omega : \omega \in F\}$ are in the simplicial complex. The method works, rather, in reverse by describing which intersections are not in the complex. Each intersection is defined by an index set $I \subseteq \{1, \ldots, N\}$

$$Q_{\vee_{i \in I} \omega^{(i)}} = \bigcap_{i \in I} Q_{\omega^{(i)}}$$

Definition 49 *For an index set I and $j \notin I$ we say that j covers I if*

(i) $j < i$ for all $i \in I$

(ii) $Q_{\omega^{(j)}} \supseteq Q_{\vee_{i \in I} \omega^{(i)}}$

Note that *(ii)* is equivalent to saying

$$\omega^{(j)} \leq \vee_{i \in I} \omega^{(i)}$$

Definition 50 *An index set I is covered if there is a subset I' of I (possibly I itself) such that there is some $j \notin I'$ which covers I'.*

Now define $\mathcal{S}$ as the set of non-covered index sets J. The main result is that these form a discrete tube.

Theorem 39 *Select a listing of the minimal failure outcomes:*

$$F = \left\{\omega^{(1)}, \ldots, \omega^{(N)}\right\}$$

and let $\mathcal{Q} = \{Q_\omega : \omega \in F\}$ be the corresponding individual failure events. Let $\mathcal{S}$ be the set of non-covered index sets $J \subseteq \{1, \ldots, N\}$. Then, $\{\mathcal{Q}, \mathcal{I}, \mathcal{S}\}$ forms an abstract tube ($\mathcal{I} = \{1, \ldots, N\}$). Moreover the depth of the tube is less than or equal to d, the number of components.

Proof. The proof is omitted but is an adaptation of the proof in Naiman and Wynn (2000). See also Giglio, Naiman and Wynn (2000). □

Theorem 39 can then be immediately applied to give the following result for coherent systems.

Theorem 40 *For a coherent system D_{2^d}, let $F = \{\omega^{(1)}, \ldots, \omega^{(n)}\}$ be the set of minimal failure outcomes, indexed in some order. Let $\{\mathcal{Q}, \mathcal{I}, \mathcal{S}\}$ be the tube described in Theorem 39. Let D be the (maximal) failure event,* *then*

$$f_D(x) = \sum_{r=1}^{d}(-1)^{r-1}\sum_{\substack{\#I=r \\ I\in\mathcal{S}}}\prod_{w\in I}q_w(x)$$

$$\leq \sum_{r=1}^{s}(-1)^{r-1}\sum_{\substack{\#I=r \\ I\in\mathcal{S}}}\prod_{w\in I}q_w(x) \text{ (s odd)}$$

$$\geq \sum_{r=1}^{s}(-1)^{r-1}\sum_{\substack{\#I=r \\ I\in\mathcal{S}}}\prod_{w\in I}q_w(x) \text{ (s even)}$$

$(1 \leq s < d)$.

Example 59 Consider another example in five dimensions with minimal failure events

$$\omega^{(1)} = 11000, \omega^{(2)} = 00111, \omega^{(3)} = 10101, \omega^{(4)} = 01110$$

Computing the intersections via the $\omega^{(i)}$, we obtain

$\omega^{(1)} \vee \omega^{(2)}$	$= 11111$	not covered
$\omega^{(1)} \vee \omega^{(3)}$	$= 11101$	not covered
$\omega^{(1)} \vee \omega^{(4)}$	$= 11110$	not covered
$\omega^{(2)} \vee \omega^{(3)}$	$= 10111$	not covered
$\omega^{(2)} \vee \omega^{(4)}$	$= 01111$	not covered
$\omega^{(3)} \vee \omega^{(4)}$	$= 11111$	covered
$\omega^{(1)} \vee \omega^{(2)} \vee \omega^{(3)}$	$= 11111$	not covered
$\omega^{(1)} \vee \omega^{(2)} \vee \omega^{(4)}$	$= 11111$	not covered
$\omega^{(1)} \vee \omega^{(3)} \vee \omega^{(4)}$	$= 11111$	covered
$\omega^{(2)} \vee \omega^{(3)} \vee \omega^{(4)}$	$= 11111$	covered
$\omega^{(1)} \vee \omega^{(2)} \vee \omega^{(3)} \vee \omega^{(4)}$	$= 11111$	covered

The simplicial complex, in terms of the indices, is

$$\{1\}, \quad \{2\}, \quad \{3\}, \quad \{4\},$$
$$\{1,2\}, \quad \{1,3\}, \quad \{1,4\}, \quad \{2,3\}, \quad \{2,4\},$$
$$\{1,2,3\}, \quad \{1,2,4\}$$

This gives an improvement over the classical inclusion-exclusion lemma at depths two and three because $\{3,4\}$ is omitted at depth two and $\{1,3,4\}$ and $\{2,3,4\}$ are omitted at depth three. The depth of the tube is three as opposed to four for the inclusion-exclusion lemma. The simplicial complex is given pictorially by Figure 4.2.

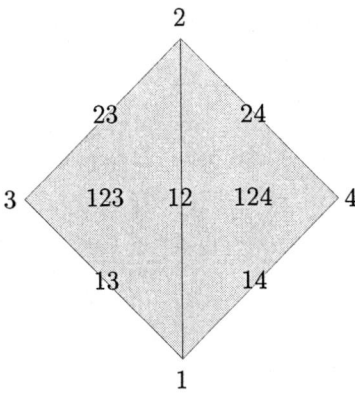

Figure 4.2 *A simplicial complex.*

4.5.1 Failure/non-failure duality and Gröbner bases

Because the failure and non-failure events for a coherent system are both minimal fan echelon designs it is natural to conjecture a duality involving Gröbner bases.

Consider the simple two-component system with single failure outcome 11. The non-failure event is $\{00, 10, 01\}$. As a zero-dimensional point set this has the Gröbner basis $\{x_1^2 - x_1, x_2^2 - x_2, x_1x_2\}$. Note that x_1x_2 has index vector 11, which is precisely the failure event. Now interchange 0 and 1 so that $11 \mapsto 00$. The Gröbner basis for the single point 00 is $\langle x_1, x_2 \rangle$ whose elements have index set 10 and 01, which are the transformed version of the minimal non-failure events 01 and 10, respectively. Notice that we do not use the quadratic elements $x_i^2 - x_i$.

We collect this duality together as a theorem, but omit the proof. To aid the statement we refer to the $0 \leftrightarrow 1$ interchange as "flipping".

Theorem 41 *Let F and $G = \{0,1\}^d \setminus F$ be respectively the failure and non-failure sets for a (binary) coherent system. Let $\tilde{F}$ and $\tilde{G}$ be their flipped version. Let $L(S)$ be the leading terms of the reduced Gröbner basis of a subset $S \subseteq \{0,1\}^d$ excluding quadratic terms. Then*

$$L(G) = F^\star$$
$$L(\tilde{F}) = \tilde{G}^\star$$

where $F^\star$ and $\tilde{G}^\star$ are the minimum outcomes in each case.

Example 60 [Continuation of Example 58]. Either by taking complements or noting that the minimal paths in Figure 4.2 are 14, 25, 135 and 234, the minimal non-failure events are 01101, 10110 01010 and 10001 and the full non-failure event set is the complement set of Table 4.1:

```
01101   10110
10100   10010   10001   01100
01010   01001   00110   00101
10000   01000   00100   00010   00001
00000
```

Moreover, $L(G)$ for the above point set is x_1x_2, x_4x_5, $x_1x_3x_5$ and $x_2x_3x_4$ confirming the result that $L(G) = F^\star$. We end this example by computing a tube simplicial complex for the non-failure events. There are two options to change the tube construction in Definition 49 by replacing $\wedge$ by $\vee$ or simply applying the construction to $\tilde{F}$. Let us do the latter. Thus

$$\tilde{F}^\star = \{10010, 01001, 10101, 01110\}.$$

Using Definition 49 in the order shown, we have

$$\begin{aligned}
\omega^{(1)} \vee \omega^{(2)} &= 11011 \quad \text{not covered} \\
\omega^{(1)} \vee \omega^{(3)} &= 10111 \quad \text{not covered} \\
\omega^{(1)} \vee \omega^{(4)} &= 11110 \quad \text{not covered} \\
\omega^{(2)} \vee \omega^{(3)} &= 11101 \quad \text{not covered} \\
\omega^{(2)} \vee \omega^{(4)} &= 01111 \quad \text{not covered} \\
\omega^{(3)} \vee \omega^{(4)} &= 11111 \quad \text{covered} \\
\omega^{(1)} \vee \omega^{(2)} \vee \omega^{(3)} &= 11111 \quad \text{not covered} \\
\omega^{(1)} \vee \omega^{(2)} \vee \omega^{(4)} &= 11111 \quad \text{not covered} \\
\omega^{(1)} \vee \omega^{(3)} \vee \omega^{(4)} &= 11111 \quad \text{covered} \\
\omega^{(2)} \vee \omega^{(3)} \vee \omega^{(4)} &= 11111 \quad \text{covered} \\
\omega^{(1)} \vee \omega^{(2)} \vee \omega^{(3)} \vee \omega^{(4)} &= 11111 \quad \text{covered}
\end{aligned}$$

This gives the simplicial complex (in terms of the index set for intersections):

$$\begin{aligned}
&\{1\}, \quad \{2\}, \quad \{3\}, \quad \{4\}, \\
&\{1,2\}, \quad \{1,3\}, \quad \{1,4\}, \quad \{2,3\}, \quad \{2,4\} \\
&\{1,2,3\}, \quad \{1,2,4\}
\end{aligned}$$

We may obtain alternative inequalities for system failure by setting out the bounds for non-failure and using probability{system failure} = 1 − probability{system non-failure}.

4.6 Two-level factorial design: contrasts and orthogonality

If the binary experimental design is treated from the point of view of classical design rather than its adaptation for logic and reliability, then somewhat different issues arise. Foremost among these is orthogonality and its relationship to the underlying group structure. The following development is taken from Fontana, Pistone and Rogantin (1997) and Fontana, Pistone and Rogantin (2000).

It is convenient to switch to a different binary coding $\{-1, 1\}$ and the unit hyper-cube $\{-1, 1\}^d$. All the Gröbner basis theory still applies in the sense that the natural recoding $1 \to 1$ and $0 \to -1$ preserves $\text{Est}_\tau(D)$. This is because the Gröbner basis for the d-dimensional full factorial design over $\{-1, 1\}$ is
$$\langle x_i^2 - 1 : i = 1, \ldots, d \rangle$$
with respect to any term-ordering.

In developing the application to 2^d factorial design we shall make special use of indicator functions. Here is a list of properties. Recall Definition 41.

1. The polynomial indicator function on $\{-1, 1\}^d$ of a single point a is
$$f_a(x) = \prod_{i: a_i = 1} \frac{1 + x_i}{2} \prod_{i: a_i = -1} \frac{1 - x_i}{2} = \prod_{i=1}^d \frac{1 + a_i x_i}{2}$$
and of a design D is
$$f_D(x) = \sum_{a \in D} f_a(x)$$

2. For a fraction D of D_{2^d} the design ideal is
$$\text{Ideal}(D) = \langle x_1^2 - 1, \ldots, x_d^2 - 1, 1 - f_D(x) \rangle$$
In general this representation is not a Gröbner basis.

3. We can characterize $f_D(x)$ also as
$$\begin{cases} \text{Ideal}(D) = \langle x_1^2 - 1, \ldots, x_d^2 - 1, 1 - f_D(x) \rangle \\ f_D^2 - f_D \in \langle x_1^2 - 1, \ldots, x_d^2 - 1 \rangle \end{cases}$$

The theory of 2^d fractions is well managed if we essentially restrict all the statements to calculations of y-values on $\{-1, 1\}^d$. Thus for any function from the factorial design D_{2^d} to a field k we immediately replace it by its polynomial interpolator
$$f_{D_{2^d}}(x) = \sum_{\alpha \in D_{2^d}} \theta_\alpha x^\alpha$$
We now define an "expectation operator" $E(\cdot)$ which simply averages quantities over $\{-1, 1\}^d$.

Definition 51 *Let f be a function from D_{2^d} to a field k. Then, the* mean value *of f is denoted by $E(f)$ and is defined by*
$$E(f) = \frac{1}{2^d} \sum_{x \in \{-1, 1\}^d} f(x)$$

This operator is useful for formalizing the idea of orthogonality.

Definition 52 *Let f, g be a functions from D_{2^d} to a field k.*

1. A contrast *on D_{2^d} is a function (response) f such that*

$$E(f) = 0$$

2. Two functions, f and g, are orthogonal *on D_{2^d} if*

$$E(fg) = 0$$

For a design $D \subseteq \{-1, 1\}^d$, the above definition extends to contrasts and orthogonality with respect to D if the expectations are restricted to D.

Definition 53 *Let D be a fraction of the 2^d full factorial design and let f, g be functions on D. The* mean value *of f is defined as*

$$E_D(f) = \frac{1}{\#D} \sum_{x \in D} f(x)$$

where $\#D$ is the number of elements of D. The function f is a contrast *on D if*

$$E_D(f) = 0$$

The functions f and g are said to be orthogonal *on D if their product is a contrast*

$$E_D(fg) = 0$$

We investigate when two contrasts f and g are orthogonal with respect to a fraction D. The first point is to express the indicator function of the fraction D in terms of the monomials in Est $= \{x^\alpha : \alpha \in L_{2^d}\}$, in other words, expand $f_D(x)$ into multi-linear monomials, independent of the term-ordering. Let this be

$$f_D(x) = \sum_{\alpha \in L_{2^d}} b_\alpha x^\alpha \qquad (4.4)$$

One advantage of the $\{-1, 1\}$ coding is the special role played by the symmetric difference of index sets. As we are only interested in multi-linear terms, we identify a monomial x^α with the vector α where $\alpha_i = 0$ or 1 and in turn α can be identified with the set recording the position of the ones in α. For example, for $\alpha = (1, 0, 1, 0)$, the corresponding set is $\{1, 3\}$ and for $\beta = (0, 1, 1, 1)$, it is $\{2, 3, 4\}$. Then, we can define

$$\alpha \triangle \beta = (\alpha \cup \beta) \setminus (\alpha \cap \beta)$$

with reference to the set representation. For example, for $\alpha = (1, 0, 1, 0)$ and $\beta = (0, 1, 1, 1)$ we have $\alpha \triangle \beta = (1, 1, 0, 1)$. It is important to note that thus the monomial $x_1 x_2 x_4$ is the product of the monomials $x_1 x_3$ and $x_2 x_3 x_4$ in the Est representation of the quotient space $k[x_1, \ldots, x_4]/\text{Ideal}(D_{2^4})$.

Theorem 42 *Let x^α, x^β and x^γ be multi-linear monomial functions over*

the full factorial design D_{2^d}, that is $\alpha_i, \beta_i, \gamma_i = 0$ or 1 for $i = 1, \ldots, d$. Then
$$\begin{cases} E(x^\alpha) = 0 & \text{for } x^\alpha \neq 1 \\ E(x^\alpha x^\beta) = 0 & \text{for } x^\alpha \neq x^\beta \end{cases}$$
that is all the multi-linear terms (but 1) are contrasts and any pair of distinct multi-linear terms are orthogonal with respect to a full factorial design.

Now let D be a fraction of D_{2^d} and let f_D be the indicator function of D as in Equation (4.4). Then

1. x^α is a contrast on D if and only if $b_\alpha = 0$
2. x^α and x^β are orthogonal on D if and only if $b_{\alpha \triangle \beta} = 0$
3. Let x^α be a contrast. For any x^β and x^γ such that $\alpha = \beta \triangle \gamma$, x^β is orthogonal to x^γ over D.

Moreover, we have
$$\begin{cases} E_D(x^\alpha) = \dfrac{2^d}{\#D} b_\alpha \\ E_D(x^\alpha x^\beta) = \dfrac{2^d}{\#D} b_{\alpha \triangle \beta} \end{cases}$$

Proof. See Fontana, Pistone and Rogantin (1997). A useful relation used in the proof is $x^\alpha x^\beta = x^{\alpha \triangle \beta}$ over D_{2^d}. □

Given a term-ordering τ, the aim is to construct a maximal (in number) set of mutually orthogonal contrasts among the x^α, for $x^\alpha \in \text{Est}_\tau(D)$. According to Theorem 42, given the indicator function f_D for the design D as in Equation (4.4), we may simply inspect the coefficients b_α. If $\{x^\alpha : \alpha \in L'\}$ is a set of mutually orthogonal contrasts, then any model employing all (or some) of these terms is
$$y = \sum_{\alpha \in L'} \theta_\alpha x^\alpha + \varepsilon$$
and the least-square estimators of the θ_α are simply
$$\hat{\theta}_\alpha = E_D(x^\alpha) Y(x) \qquad \alpha \in L'$$
where $Y(x)$ is the interpolator of the observed values at the design points. These estimators are also sometimes referred to as contrasts.

Example 61 Let D be the fraction of D_{2^5} in Table 4.2. The corresponding indicator function is
$$f_D(x) = \frac{1}{2} + \frac{1}{4} x_1 x_2 x_4 - \frac{1}{4} x_1 x_2 x_3 + \frac{1}{4} x_1 x_2 x_4 x_5 + \frac{1}{4} x_1 x_2 x_3 x_5$$

Table 4.2 *A fraction of the five-dimensional full factorial design.*

1	1	1	1	1
1	1	−1	1	1
1	1	−1	1	−1
1	1	−1	−1	−1
1	−1	1	1	−1
1	−1	1	−1	1
1	−1	1	−1	−1
1	−1	−1	−1	1
−1	1	1	1	−1
−1	1	1	−1	1
−1	1	1	−1	−1
−1	1	−1	−1	1
−1	−1	1	1	1
−1	−1	−1	1	1
−1	−1	−1	1	−1
−1	−1	−1	−1	−1

With respect to the term-ordering tdeg, we have
Est(D) = { 1,
x_1, x_2, x_3, x_4, x_5,
x_1x_3, x_1x_4, x_1x_5, x_2x_3, x_2x_4, x_2x_5,
x_3x_4, x_3x_5, x_4x_5, $x_3x_4x_5$ }

and a maximal subset of mutually orthogonal monomials is

L' = { 1, x_1, x_2, x_3, x_4, x_5,
x_1x_5, x_2x_5, x_3x_4, x_3x_5, x_4x_5,
$x_3x_4x_5$ }

CHAPTER 5

Probability

The presentation in this book of the application of commutative algebra in probability and statistics has proceeded in two basic stages. In Chapters 2, 3 and 4, the emphasis is on the application of algebraic aspects to experimental design. In the present chapter, algebraic constructions are carried out for discrete finite probability distributions, replacing the design, as a set of discrete points, by the support of a discrete distribution.

The algebraic encoding of the basic setting of probability and statistics in the case of a finite sample space is carried out in a sequence of steps.

Level 1. Let k be a field of constants, typically $k = \mathbb{Q}$. Finite subsets in k^d are described as zero-dimensional varieties, i.e. as the set of solutions of a system of polynomial equations in d indeterminates. Such a subset represents the sample space, or support, directly or after a suitable coding.

Level 2. Let $\mathcal{K}$ be an extension of the basic field k in Level 1, typically $\mathcal{K} = \mathbb{R}$. The $\mathcal{K}$-valued random variables over a finite support form a ring and are described as the quotient ring of the polynomial ring with coefficients in $\mathcal{K}$ and d indeterminates by the variety (or more exactly its ideal) in Level 1. The monomial basis of this space does not depend on the extension field.

Level 3. Probabilities and statistical models are introduced using one of a class of possible geometries on the set of probabilities, e.g., densities, square roots of densities, exponential models. An algebraic representation exists for each case.

Level 4. For special types of support (lattice), a further level of algebraisation is possible. In the non-lattice case, a differential-algebraic setting is useful.

Level 3 mentions "geometry", meaning that the actual construction of probability and later probability models encourages the introduction of extra structure. That is, commutative algebra is not enough. More precisely, as the probability is associated to expected values, hence to a scalar product, we put a Euclidean structure on the ring of random variables. Such a structure can be naturally derived from a special representation of the ring of random variables as a (commutative) ring of matrices.

5.1 Random variables on a finite support

In the present chapter, a finite subset of the affine space k^d, namely the design in the previous chapters

$$D = \{a(i) \in k^d : i = 1, \ldots, n\}$$

is called *support* and represents the sample space, that is support, of a discrete finite probability distribution. We often write $a_D := \{a : a \in D\}$ and consider a_D as an ordered list for a fixed but otherwise arbitrary total ordering. We recall its algebraic representation from the previous chapters. Let

$$\text{Ideal}(D) = \{f \in k[x] : \text{for all } a \in D, \ f(a) = 0\}$$

be the ideal associated to D. It will be called the *support ideal*. It is a radical ideal and defines uniquely the support D as its variety. As in the design case, given the support D and a term-ordering τ, there is a unique set of terms which forms a vector space basis of the quotient space $k[x]/\text{Ideal}(D)$

$$\text{Est}_\tau(D) = \{x^\alpha : \alpha \in L\}$$

(see Section 3.5).

In general (see Section 2.2) any function Y from D to $\mathcal{K}$, an extension field of k, can be represented as

$$Y = \sum_{\alpha \in L} c_\alpha X^\alpha \qquad (5.1)$$

with $c_\alpha \in \mathcal{K}$. Here X^α with capital X denotes the function

$$X^\alpha : D \longrightarrow k \subseteq \mathcal{K}$$

while the small case letter x is used for the indeterminate. That is $\{X^\alpha\}_{\alpha \in L}$ is a basis of the vector space of $\mathcal{K}$-valued functions on D. Equation (5.1) above is related to the normal form of polynomial interpolators of $Y = \{Y(a) : a \in D\}$.

Note that if the field k has characteristic zero, that is it contains the rational numbers, and D is a finite set of points in k^d, then a Gröbner basis of $k[x_1, \ldots, x_d]/\text{Ideal}(D)$ is also a Gröbner basis of $\mathcal{K}[x_1, \ldots, x_d]/\text{Ideal}(D)$, for all extension fields $\mathcal{K}$ of k. This can be easily seen as the operations needed to compute the Gröbner basis of $\text{Ideal}(D)$ in $\mathcal{K}[x_1, \ldots, x_d]$ are actually performed over $k[x_1, \ldots, x_d]$. On this point see also Section 5.2.

Definition 54 *Let D be a finite set of distinct points in k^d and let $\mathcal{K}$ be an extension field of k. The vector space of all functions from D to $\mathcal{K}$ is indicated as $\mathcal{L}(D, \mathcal{K})$, briefly $\mathcal{L}(D)$ or $\mathcal{L}$.*

As in the design case, note that the list L depends on τ and D and should therefore always be taken in context. Note also that $\text{Est}_\tau(D)$ is the complementary set of the set of leading terms of the support ideal $I(D)$. We consider L as a list ordered according to τ. From the previous theory (see in particular Section 2.10), we know that the following ring isomorphism holds

$$\mathcal{L}(D,\mathcal{K}) \sim \mathcal{K}[x_1,\ldots,x_d]/\operatorname{Ideal}(D)$$

5.2 The ring of random variables

We now consider the algebraisation of the event space and subsequently the space of random variables.

Subsets of designs are called fractions in experimental design theory (see Definition 39). In the context of discrete random variables we have the following definition, which we saw briefly in Chapter 4.

Definition 55 *An event, A is a subset of a sample space D.*

As in Definition 41 an *event* A in the support D can be identified by its indicator function f_A. The indicator function has a polynomial representation on the support of type (5.1)

$$f_A = \sum_{\alpha \in L} c_\alpha X^\alpha \qquad (5.2)$$

If $[a(1),\ldots,a(N)]$ is the ordered list of the sample points, then Equation (5.2) evaluated at $a(i)$, $i = 1,\ldots,N$, induces the following system of linear equations

$$\begin{cases} f_A(a(1)) = \sum_{\alpha \in L} c_\alpha a(1)^\alpha \\ \quad\vdots \\ f_A(a(N)) = \sum_{\alpha \in L} c_\alpha a(n)^\alpha \end{cases}$$

As in design theory we define the *support (or design) matrix* (see Definition 26)

$$Z = [a(i)^\alpha]_{i=1,\ldots,N;\alpha \in L}$$

so that the indicator function of the sample point $a(i)$ in Equation (5.2) has entries $c_{\alpha,i}$ given by the vectors $[c_i] = [c_{\alpha,i}]_{\alpha \in L}$, thus

$$[c_i] = Z^{-1}[e_i]$$

where we have performed the identification of functions with column vectors, e_i is the i-th element of the canonical basis of the vector space k^n and $[e_i]$ is the corresponding column vector. Note that Z is invertible as $\operatorname{Est}_\tau(D)$ is a vector-space basis. Note also the formula

$$Z^{-1} = [c_1,\ldots,c_N]$$

A *random variable* Y on the support D with values in the field $\mathcal{K}$, has the form of Equation (5.1), with coefficients $c_\alpha \in \mathcal{K}$. In fact, it can be written as $\mathcal{K}$-linear combinations of the indicator functions of the points of the support. If $[y_D] = [Y(a(1)), \ldots, Y(a(N))]$ is the vector of the values of Y and $[c_L] = [c_\alpha : \alpha \in L]$ is the vector of coefficients in Equation (5.1), then the equation

$$[c_L] = Z^{-1}[y_D] \qquad (5.3)$$

and (5.1) are two equivalent ways to describe a generic random variable on D.

The following theorem describes the ring structure of $\mathcal{L}(D)$.

Theorem 43 *The multiplication of X^α and X^β is computed as*

$$X^\alpha X^\beta = \operatorname{Rem}_\tau \left(X^{\alpha+\beta} \right)$$
$$= \sum_{\gamma \in L} r(\alpha+\beta,\gamma) X^\gamma$$

where $\operatorname{Rem}_\tau(f(X))$ means the random variable $\tilde{f}(X)$ derived from $\tilde{f}(x) = \operatorname{Rem}_\tau(f(x))$ by substituting the indeterminate x with the random variable X, and the normal form is computed with respect to $\operatorname{Ideal}(D)$.

The following definition has an important computational outcome.

Definition 56 *Define the multi-array*

$$R = [R(\alpha,\beta,\gamma)]_{\alpha,\beta,\gamma \in L} = [r(\alpha+\beta,\gamma)]_{\alpha,\beta,\gamma \in L}$$

and the list of $N \times N$ matrices

$$R(\beta) = [r(\alpha+\beta,\gamma)]_{\gamma,\alpha \in L} \qquad \beta \in L$$

whose elements are computed from the normal form operation. Sometimes we write $R(L)$ to stress the monomial basis considered.

The entries $r(\delta,\gamma)$, $\delta \in L+L$ and $\gamma \in L$ are a basic computational object. For an example, see Example 63. We assume from now on that they have been computed.

Example 62 *From the definition it follows that $R(0)$ is the identity matrix and that $R(\alpha,\beta,\gamma) = R(\alpha+1,\beta-1,\gamma)$. There is much symmetry in the multi-array R. Note moreover that not all of $L+L$ is actually needed, but by recursion it is sufficient to consider terms in L and $x^\alpha x_j$ for all indeterminates x_j.*

Using the previous representation, the product of two random variables has the following form

$$YZ = \left(\sum_{\alpha \in L} c_\alpha X^\alpha \right) \left(\sum_{\beta \in L} d_\beta X^\beta \right)$$
$$= \sum_{\gamma \in L} \left(\sum_{\alpha,\beta \in L} c_\alpha d_\beta r(\alpha+\beta,\gamma) \right) X^\gamma \qquad (5.4)$$
$$= \sum_{\gamma \in L} [c]_L^t \, [r(\alpha+\beta,\gamma]_{\alpha,\beta \in L} \, [d]_L X^\gamma$$

5.3 Matrix representation of $\mathcal{L}(D, \mathcal{K})$

The following theorem states that the ring of random variables $\mathcal{L}$ is a vector space with a monomial basis. This allows a matrix representation of the multiplication operator.

Theorem 44 *Consider the k-linear mapping*

$$\begin{aligned} T_\beta : \mathcal{L}(D) &\longrightarrow \mathcal{L}(D) \\ Y &\mapsto X^\beta Y \end{aligned}$$

where $\beta \in L$. The linear operator T_β is represented on the monomial basis as

$$\begin{aligned} X^\beta Y &= X^\beta \left(\sum_{\alpha \in L} c_\alpha X^\alpha \right) \\ &= \sum_{\alpha \in L} c_\alpha X^{\alpha + \beta} \\ &= \sum_{\gamma \in L} \left(\sum_{\alpha \in L} r(\alpha + \beta, \gamma) c_\alpha \right) X^\gamma \end{aligned} \quad (5.5)$$

If f_a is the indicator function of a point a in the support D, then f_a is an eigenfunction of T_β and the corresponding eigenvalue is a^β.

Theorem 44 above shows that the linear mapping T_β is represented by the matrix $R(\beta)$ in Definition 56. The set of matrices

$$\sum c_\beta R(\beta) \qquad c_\beta \in \mathcal{K}$$

is a representation of $\mathcal{L}(D, \mathcal{K})$ into the non-commutative ring of matrices $\mathcal{K}^{N,N}$ as a commutative subring of matrices. By substituting the indicator function $Y = f_a$ in Equation (5.5) and equating coefficients, we see that

$$[c_i] = Z^{-1}[e_i] \implies \left[\sum_{\alpha \in L} r(\alpha + \beta, \gamma) c_{\alpha,i} \right]_{\gamma \in L} = a^\beta [c_i] \quad (5.6)$$

Using the matrices $R(\beta)$, then Equation (5.6) gives

$$R(\beta) Z^{-1} = Z^{-1} \operatorname{diag}\left(a^\beta : a \in D \right), \qquad \beta \in L$$

hence

$$R(\beta) = Z^{-1} \operatorname{diag}\left(a^\beta : a \in D \right) Z, \qquad \beta \in L \quad (5.7)$$

In particular, equating traces in Equation (5.7) we obtain

$$\operatorname{Tr} R(\beta) = \sum_{a \in D} a^\beta, \qquad \beta \in L \quad (5.8)$$

Example 63 Consider a three-way layout (contingency table) with $3 \times 3 \times 3$ cells. Assume that the cells are coded $1, 2, 3$ and cells (i, i, i), $i = 1, 2, 3$ are not allowed. These are sometimes referred to as structural zeros. The corresponding polynomial ideal is obtained by elimination of t_1 and t_2 from the system

$$\begin{cases} (x_1-1)(x_1-2)(x_1-3), \\ (x_2-1)(x_2-2)(x_2-3), \\ (x_3-1)(x_3-2)(x_3-3), \\ (x_1-x_2)t_1+(x_1-x_3)t_2-1 \end{cases} \quad (5.9)$$

giving for $\mathtt{tdeg}(x_1 \succ x_2 \succ x_3)$

$$\begin{cases} (x_1-1)(x_1-2)(x_1-3), \\ (x_2-1)(x_2-2)(x_2-3), \\ (x_3-1)(x_3-2)(x_3-3), \\ x_1^2x_2^2+x_1^2x_2x_3+x_1x_2^2x_3+x_1^2x_3^2+x_1x_2x_3^2+x_2^2x_3^2-6x_1^2x_2-6x_1x_2^2 \\ \quad -6x_1^2x_3-12x_1x_2x_3-6x_2^2x_3-6x_1x_3^2-6x_2x_3^2+11x_1^2+36x_1x_2 \\ \quad +11x_2^2+36x_1x_3+36x_2x_3+11x_3^2-60x_1-60x_2-60x_3+85 \end{cases}$$

The corresponding vector space basis for $\mathcal{L}(D)$ is

$$\begin{array}{llllll}
1 & x_1 & x_2 & x_3 & & \\
x_1^2 & x_1x_2 & x_1x_3 & x_2^2 & x_2x_3 & x_3^2 \\
x_1^2x_2 & x_1^2x_3 & x_1x_2^2 & x_1x_2x_3 & x_1x_3^2 & x_2^2x_3 & x_2x_3^2 \\
x_1^2x_2x_3 & x_1^2x_3^2 & x_1x_2^2x_3 & x_1x_2x_3^2 & x_2^2x_3^2 & & \\
x_1^2x_2x_3^2 & x_1x_2^2x_3^2 & & & & &
\end{array}$$

Note that the missing terms with respect to a full 3^3 set of points are $x_1^2x_2^2, x_1^2x_2^2x_3$ and $x_1^2x_2^2x_3^2$. Note that the fan for the above ideal consists of the six leaves corresponding to the permutation of indexes $1, 2$ and 3.

The uniform probability over D is simply given by the constant polynomial

$$P_0(x_1, x_2, x_3) = \frac{1}{24} \in \mathcal{L}(D)$$

With a program in CoCoA, we computed the matrices $R(\beta)$ for this example and found the trace vector in Equation 5.10, which is given below, in the same ordering as the vector space basis above

$$\begin{aligned} \mathrm{Tr}R(\beta) = (\quad & 24, \quad 48, \quad 48, \quad 48, \\ & 112, \quad 94, \quad 94, \quad 112, \quad 94, \quad 112, \\ & 216, \quad 216, \quad 216, \quad 180, \quad 216, \quad 216, \quad 216, \quad\quad (5.10)\\ & 490, \quad 406, \quad 406, \quad 406, \quad 490, \\ & 75, \quad 75 \quad) \end{aligned}$$

Example 64 Consider the fraction of a 3^3-full factorial structure with $x \leq y$ and $x \leq z$, that is the 14 points (x, y, z)

$$\begin{array}{lllll}
(1,1,1), & (1,2,1), & (1,3,1), & (1,1,2), & (1,2,2), \\
(1,3,2), & (1,1,3), & (1,2,3), & (1,3,3), & \\
(2,2,2), & (2,3,2), & (2,2,3), & (2,3,3), & \\
(3,3,3) & & & &
\end{array}$$

The above is an echelon design and its reduced Gröbner bases is given by the following polynomials

$$(x_1 - 1)(x_1 - 2)(x_1 - 3)$$
$$(x_2 - 1)(x_2 - 2)(x_2 - 3)$$
$$(x_3 - 1)(x_3 - 2)(x_3 - 3)$$
$$(x_2 - 3)(x_1 - 1)(x_1 - 2)$$
$$(x_3 - 3)(x_1 - 1)(x_1 - 2)$$
$$(x_2 - 2)(x_2 - 3)(x_1 - 1)$$
$$(x_3 - 2)(x_3 - 3)(x_1 - 1)$$

and the identifiable terms are

$$\begin{array}{cccccc} 1 & x_1 & x_2 & x_3 & & \\ x_1^2 & x_1 x_2 & x_1 x_3 & x_2^2 & x_2 x_3 & x_3^2 \\ x_1 x_2 x_3 & x_2^2 x_3 & x_2 x_3^2 & x_2^2 x_3^2 & & \end{array}$$

The trace vector ordered as the set of terms above is

$$\mathrm{Tr} R(\beta) = (\begin{array}{cccccc} 14 & 20 & 31 & 31 & & \\ 34 & 47 & 47 & 77 & 70 & 77 \\ 113 & 176 & 176 & 446 & & \end{array}) \quad (5.11)$$

5.4 Uniform probability

Let us consider the uniform distribution $P_0(\cdot)$ on D and the corresponding expectation $E_0(\cdot)$ on $\mathcal{L}(D)$. The uniform distribution assigns equal probability to the elementary events. As all random variables are of the form (5.1), the expected value of a random variable Y can be defined as follows.

Definition 57 *The expectation of a discrete random function on $\mathcal{L}(D)$, $Y = \sum_{\alpha \in L} c_\alpha X^\alpha$ with respect to the uniform distribution is*

$$E_0(Y) = \sum_{\alpha \in L} c_\alpha E_0(X^\alpha) \quad (5.12)$$

The basic moments *are defined as*

$$m_\alpha = E_0(X^\alpha), \quad \alpha \in L$$

Note that from Equation 5.12, we obtain an instance of moment confounding or aliasing in that for all β

$$E_0(X^\beta) = \sum_{\alpha \in L} r(\beta, \alpha) E_0(X^\alpha)$$

That is $E_0(X^\beta)$ is expressed as a finite sum of lower-order moments. We return to study such issues in more detail in Section 5.9.

Theorem 45 *With the notation of Definition 57, the basic moments with respect to the uniform distribution are computed as*

$$m_\alpha = \frac{1}{\#D} \mathrm{Tr} R(\alpha)$$

where $\#D = \#L$ *is the number of distinct points in D.*

Proof. The proof follows from Equation 5.8 and the definition of expectation with respect to the uniform distribution. □

Theorem 45 gives a formula relating the moments to the Gröbner basis which we shall use again.

Expectation computation can be expressed through moments.

Theorem 46 *With the notation of Definition 57, let $[m_L] = [m_\alpha]_{\alpha \in L}$ and the vector $[y]$ contains the values of Y on D, then*

$$\mathrm{E}_0(Y) = [m]^t[c] = [m]^t Z^{-1}[y] \qquad (5.13)$$

Proof. This follows from Equation (5.3) and Theorem 45 above. □

Example 65 [Continuation of Example 63] The basic moments are simply obtained by dividing the trace vector by 24

$$[m]_L = (\quad 1,\quad 2,\quad 2,\quad 2,$$
$$14/3,\quad 47/12,\quad 47/12,\quad 14/3,\quad 47/12,\quad 14/3,$$
$$9,\quad 9,\quad 9,\quad 180,\quad 9,\quad 9,\quad 9,$$
$$245/12,\quad 203/12,\quad 203/12,\quad 203/12,\quad 245/12,$$
$$75/24,\quad 75/24\quad)$$

Example 66 If Y is the indicator function of the sample point $a \in D$, then its expectation is $1/N$, where N is number of points in the support. From Equation (5.13) we obtain the formula

$$\frac{1}{N} = [m]^t Z^{-1}[e_i], \qquad i = 1, \ldots, N$$

and the following identities

$$[e_i]^t (Z^{-1})^t [m] = 1/N$$
$$[e_i][e_i]^t (Z^{-1})^t [m] = [e_i]/N$$
$$\sum_{i=1}^{N} [e_i][e_i]^t (Z^{-1})^t [m] = \sum_{i=1}^{N} [e_i]/N$$
$$(Z^{-1})^t [m] = 1/N [1]_N$$
$$[m] = \frac{1}{N} Z^t [1]_N$$

where $[1]_N$ is the N-dimensional vector $[1, \ldots, 1]$.

PROBABILITY DENSITIES   103

The ring of random variables $\mathcal{L}(D)$ is endowed with a scalar product
$$\mathcal{L}(D) \times \mathcal{L}(D) \ni (Y, Z) \mapsto \mathrm{E}_0\left(YZ\right)$$
whose representation is
$$\begin{aligned}
\mathrm{E}_0\left(YZ\right) &= \mathrm{E}_0\left(\left(\sum_{\alpha \in L} a_\alpha X^\alpha\right)\left(\sum_{\beta \in L} b_\beta X^\beta\right)\right) \\
&= \sum_{\alpha, \beta \in L} a_\alpha b_\beta \mathrm{E}_0\left(X^{\alpha+\beta}\right) \\
&= \sum_{\alpha, \beta \in L} a_\alpha b_\beta \sum_{\gamma \in L} r(\alpha+\beta, \gamma) \mathrm{E}_0\left(X^\gamma\right) \\
&= \sum_{\alpha, \beta \in L} a_\alpha b_\beta \sum_{\gamma \in L} r(\alpha+\beta, \gamma) m_\gamma
\end{aligned}$$

As in Definition 56, we define a basic computational object for the computation of scalar products..

Definition 58 *Define the matrix $[Q(\alpha, \beta)]_{\alpha, \beta \in L}$ with*
$$Q(\alpha, \beta) = \mathrm{E}_0\left(X^{\alpha+\beta}\right) = \sum_{\gamma \in L} r(\alpha+\beta, \gamma) m_\gamma$$

With respect to the monomial basis of $\mathcal{L}(D)$, the scalar product is represented by the matrix Q above and also, using the vectors $[y]$ and $[z]$ of the values of Y and Z and $\mathrm{E}_0\left(YZ\right) = \frac{1}{N}[y]^t[z]$, we obtain
$$Q = \frac{1}{N} Z^t Z$$

Equation (5.8) gives also the representation
$$Q(\alpha, \beta) = \frac{1}{\#D} \sum_{\gamma, \delta \in L} r(\alpha+\beta, \gamma) r(\gamma+\delta, \delta)$$

Example 67 [Continuation of Example 63] For $x^\alpha = x_1^2 x_3$ and $x^\beta = x_1 x_2 x_3$, $Q(\alpha, \beta) = 1115/12$.

5.5 Probability densities

We now consider the case of an arbitrary probability distribution $P(\cdot)$ on D which, as a density function with respect to the uniform distribution on D, can be expressed as a member of the ring $\mathcal{L}(D)$
$$P = \sum_{\alpha \in L} \theta_\alpha X^\alpha \tag{5.14}$$

with the additional properties $P \geq 0$ on D and the normalising condition

$$\mathrm{E}_0(P) = \sum_{\alpha \in L} \theta_\alpha m_\alpha = 1 \qquad (5.15)$$

If we derive θ_0 from Equation 5.15 and substitute it in Equation 5.14, then we obtain a representation of $P(\cdot)$, including the normalising condition

$$P = 1 + \sum_{\alpha \in L_0} \theta_\alpha (X^\alpha - m_\alpha)$$

as $m_0 = 1$. The parameters θ_α, $\alpha \in L_0$ are to be chosen in the polyhedral region defined by

$$1 + \sum_{\alpha \in L_0} \theta_\alpha (a^\alpha - m_\alpha) \geq 0 \text{ for all } a \in D$$

A simple example is given by the Bernoulli $P = 1 + \theta_1 (x - 1/2)$, where $-2 \leq \theta_1 \leq 2$.

As in Section 5.4, the following definition is well posed.

Definition 59 *An element $Y = \sum_{\alpha \in L} c_\alpha X^\alpha$ of $\mathcal{L}(D)$ has expectation with respect to P given by*

$$\mathrm{E}_p(Y) = \mathrm{E}_0(PY)$$
$$= \mathrm{E}_0\left(\left(\sum_{\alpha \in L} c_\alpha X^\alpha\right)\left(\sum_{\beta \in L} \theta_\beta X^\beta\right)\right)$$
$$= \sum_{\alpha,\beta \in L} c_\alpha \theta_\beta Q(\alpha,\beta)$$

where $Q(\alpha,\beta) = \mathrm{E}_0\left(X^{\alpha+\beta}\right)$ as in Definition 58. The basic moments are

$$\mu_\alpha = \mathrm{E}_P(X^\alpha) = \sum_{\beta \in L} \theta_\beta Q(\alpha,\beta), \quad \alpha \in L$$

or $[\mu] = Q[\theta]$.

Other representations of P are possible. In Section 5.8 we shall discuss the very important exponential representation.

Consider now $S = \sqrt{P}$, the positive square root of P on D. This has its own representation

$$S = \sum_{\alpha \in L} \phi_\alpha X^\alpha$$

Now the mass unity condition is

$$\mathrm{E}_0(S^2) = 1$$

which becomes

$$E_0\left(\left(\sum_{\alpha\in L}\phi_\alpha X^\alpha\right)\left(\sum_{\beta\in L}\phi_\beta X^\beta\right)\right) = \sum_{\alpha,\beta\in L}\phi_\alpha\phi_\beta Q(\alpha,\beta) = 1$$

that is S lies on the surface of the unit sphere. Recall that $Q(\alpha,\beta) = \dfrac{1}{N}Z^t Z$ is positive definite which means that we have two contrasting encodings of the probability as linear and as spherical, respectively

$$\begin{cases} p = \sum_{\alpha\in L}\theta_\alpha x^\alpha \geq 0 & x\in D \\ \sum_{\alpha\in L}\theta_\alpha m_\alpha = 1 \end{cases}$$

and

$$\begin{cases} p = \left(\sum_{\alpha\in L}\phi_\alpha x^\alpha\right)^2 & x\in D \\ \sum_{\alpha,\beta\in L}\phi_\alpha\phi_\beta Q(\alpha,\beta) = 1 \end{cases}$$

Again, a simple example is

$$\phi_\alpha = \frac{1}{\sqrt{\sum_{\gamma,\beta\in L} Q(\gamma,\beta)}}$$

for all $\alpha \in L$.

Example 68 Consider a simple Bernoulli model. In such a case we have $D = \{0,1\}$, $\text{Ideal}(D) = \text{Ideal}(x(x-1))$ and $f(x) = p^x(1-p)^{1-x}$ (note the change in the reference measure). The two representations p, s and the logarithm $\log p$ are

$$p(x) = 1 - p + (2p-1)x$$
$$s(x) = \sqrt{1-p} + \left(\sqrt{p} - \sqrt{1-p}\right)x$$
$$\log p(x) = \log(1-p) + (\log p - \log(1-p))x$$
$$p(x) = \exp\left(\log(1-p) + \log\left(\frac{p}{1-p}\right)x\right)$$

As a more complex example, leading to the general binary case, consider two binary variables

$$\begin{cases} x_1(x_1-1) = 0 \\ x_2(x_2-1) = 0 \end{cases}$$

Thus, the set $\text{Est}_\tau(D)$ is $\{1, x_1, x_2, x_1x_2\}$ and is independent of the term-

ordering. Thus, the density is written as
$$\begin{aligned}p(x) &= p_{00} + (p_{10} - p_{00})x_1 + (p_{01} - p_{00})x_2 \\ &\quad + (p_{00} - p_{10} - p_{01} + p_{11})x_1 x_2 \\ &= \exp\left(\log p_{00} + \log\left(\frac{p_{10}}{p_{00}}\right)x_1 + \log\left(\frac{p_{01}}{p_{00}}\right)x_2 \right. \\ &\quad \left. + \log\left(\frac{p_{10}p_{01}}{p_{10}p_{11}}\right)x_1 x_2\right)\end{aligned}$$

5.6 Image probability and marginalisation

Let $Y_1, \ldots, Y_h$ be in $\mathcal{L}(D)$, the ring of functions over D, P a probability on D and write
$$Y_j = \sum_{\alpha \in L} c_{\alpha,j} X^\alpha \tag{5.16}$$
Next we compute the algebraic encoding of the density $P_{Y_1, \ldots, Y_h}$ of the joint law $Y = (Y_1, \ldots, Y_h)$.

Theorem 47 *Let $G = \{g_1, \ldots, g_k\}$ be a Gröbner basis for* $\mathrm{Ideal}(D)$ *and let $J \subseteq \mathcal{K}[y_1, \ldots, y_h, x_1, \ldots, x_d]$ be the ideal generated by*
$$\begin{cases} g_i(x) & i = 1, \ldots, k \\ y_j - \sum_{\alpha \in L} c_{\alpha,j} x^\alpha & j = 1, \ldots, h \end{cases} \tag{5.17}$$

(i) The ideal J is the ideal of the extended support of points
$$\{(a, Y(a)) : a \in D\}, \quad \text{where } Y_j = \sum_{\alpha \in L} c_{\alpha,j} X^\alpha$$

(ii) The Gröbner basis with respect to an elimination term-ordering of the y variables includes a set of polynomials f_l, $l = 1, \ldots, \bar{k}$ only in the variables y and other polynomials in the y and x variables.

(iii) The polynomials $f_l(y)$, $l = 1, \ldots, \bar{k}$ form a Gröbner basis, $G^\star$ of the image support $D^\star = Y(D)$. Let $\mathrm{Est}_\tau(D^\star) = \{y^\beta : \beta \in L^\star\}$.

(iv) The basic moments of the Y-variables are expressed as functions of the moments of the X-variables as
$$\mathrm{E}_0\left(\prod_{j=1}^h Y_j^{\beta_j} P\right) = \mathrm{E}_0\left(\prod_{j=1}^h \left(\sum_{\alpha \in L} c_{\alpha,j} X^\alpha\right)^{\beta_j} \sum_{\alpha \in L} \theta_\alpha X^\alpha\right), \quad \beta \in L^\star$$

Proof. Items (i),(ii),(iii) follow from the elimination theory of Section 2.9. To prove Item (iv), note that the vector space basis of $\mathcal{K}[y_1, \ldots, y_h]/J$ determined by $G^\star$ is $\mathrm{Est}_\tau(Y(D^\star)) = \{y^\beta : \beta \in L^\star\}$. The joint probability law of Y is known when the moments are known and the moments can be computed by elimination as follows.

The moments of the Y-variables

$$\mathrm{E}_P\left(Y^\beta\right) = \mathrm{E}_P\left(\prod_{j=1}^h Y_j^{\beta_j}\right)$$

can be reduced by moment aliasing on $D^\star$ to the basic moments for $\beta \in L^\star$ and the latter ones are computed by substitution of (5.16) as

$$\mathrm{E}_0\left(\prod_{j=1}^h Y_j^{\beta_j} P\right) = \mathrm{E}_0\left(\prod_{j=1}^h \left(\sum_{\alpha \in L} c_{\alpha,j} X^\alpha\right)^{\beta_j} \sum_{\alpha \in L} \theta_\alpha X^\alpha\right)$$

□

This solves the image problem in the sense that we are able to compute $\mathrm{E}_{Y_1,\ldots,Y_h}(F)$ as $\mathrm{E}_P(F(Y_1,\ldots,Y_h))$. Notice that

$$\mathcal{L}(Y_1,\ldots,Y_h) = \{F(Y_1,\ldots,Y_h) : F(y_1,\ldots,y_h) \in \mathcal{K}(y_1,\ldots,y_h)/\mathrm{Ideal}(D^\star)\}$$

is a sub-ring of the ring $\mathcal{L}(D)$, precisely the ring of Y-measurable random variables. The density $P^\star$ of the image measure $\mathrm{P}_{Y_1,\ldots,Y_h}(\cdot)$ is defined as the relation

$$\mathrm{E}_0(P^\star F) = \mathrm{E}_{P^\star}(F) = \mathrm{E}_P(F(Y_1,\ldots,Y_h)) = \mathrm{E}_0(PF(Y_1,\ldots,Y_h))$$

for all $F \in \mathcal{L}(D^\star)$. This is related to the computation of

$$\mathrm{E}_0(P|Y_1,\ldots,Y_h) = P^\star(Y_1,\ldots,Y_h)$$

which is the density with respect to the image of the uniform measure. Computation of conditional expectations is discussed in Section 5.7 below.

Example 69 Consider the case of the marginalisation $Y_j = X_j$ for $j = 1,\ldots,h$, $h < d$. Let τ be an elimination term-ordering for $x_1,\ldots,x_h$. Then the corresponding Gröbner basis includes polynomials $g_i(x_1,\ldots,x_h)$ for $i = 1,\ldots,k'$ and other equations in all the variables $x_1,\ldots,x_d$. Notice that in the marginalisation case we do not need to introduce the y variables.

The image support $D^\star$ is the projection of D on $x_1,\ldots,x_h$. Moreover, $\mathrm{Ideal}(D^\star)$ has Gröbner basis $\{g_i : i = 1,\ldots,k'\}$ and $\mathrm{Est}_\tau(D^\star)$ is the subset of $\mathrm{Est}_\tau(D)$ involving only the $x_1,\ldots,x_h$. Then the generic $Y \in \mathcal{L}(D^\star)$ has the form

$$Y = \sum_{\beta \in L^\star} c_\beta X^\beta$$

and note that $c_\beta = 0$ for $\beta \in L \setminus L^\star$. It follows that we can write the expectation with respect to the marginal of the P distribution as

$$\mathrm{E}_{X_1,\ldots,X_h}(Y) = \mathrm{E}_0(YP) = \sum_{\alpha \in L, \beta \in L^\star} \theta_\alpha c_\beta Q(\alpha,\beta) = [c]^t Q_{L^\star,L}[\theta]$$

We can also compute the expectation with respect to the uniform distribution on $D^\star$, $P^\star$

$$E_P(Y) = E_0^\star(YP^\star) = \sum_{\alpha,\beta \in L^\star} \theta_\alpha^\star c_\beta Q(\alpha,\beta) = [c]^t QL^\star, L^\star[\theta]^\star$$

where we wrote $P^\star = \sum_{\beta \in L^\star} \theta_\beta^\star X^\beta$. As $[c]$ is a generic vector we have

$$Q\theta = \tilde{Q}\theta^\star$$

and this defines $P^\star$.

The distribution of Y could be expressed as a density with respect to the image on $D^\star$ of the uniform probability on D. In this case it is the conditional expectation of P given Y, see Section 5.7).

Example 70 [Continuation of Example 63] The support of the marginal probability over x_1 is $D^\star = \{1,2,3\}$ with Gröbner basis equal to $\{(x_1 - 1)(x_1 - 2)(x_1 - 3)\}$ and $\text{Est}(D^\star) = \{1, x_1, x_1^2\}$. For a generic function $Y \in \mathcal{L}(D^\star)$ written as $Y = c_1 + c_1 x_1 + c_2 x_1^2$ and a generic probability over D, $P = \sum_{i=1}^{24} \theta_{\alpha_i} x^{\alpha_i}$ where α_i are ordered as in (5.10), then

$$E_P(Y) = \frac{1}{24} \sum_{i,k,h=1}^{24} z_{ih}(c_0 z_{1k} + c_1 z_{2k} + c_2 z_{3k})$$

where z_{ij} is the (i,j)-th entry of the design matrix, Z again ordered as in (5.10).

5.7 Conditional expectation

Let D, τ, $G = \{g_1(x), \ldots, g_k(x)\}$ and $\text{Est}_\tau(D) = \{x^\alpha : \alpha \in L\}$ be as usual. Let

$$Z = \sum_{\alpha \in L} b_\alpha X^\alpha, \quad Y_j = \sum_{\alpha \in L} c_{\alpha,j} X^\alpha, \quad j = 1, \ldots, l$$

be $l+1$ random variables on D. We are interested in computing the conditional expectation

$$\hat{Z} = E_P(Z|Y_1, \ldots, Y_l) = f(Y_1, \ldots, Y_l)$$

First, define a general function of the type $g(Y_1, \ldots, Y_l)$ on D. All such functions form a sub-ring of $\mathcal{L}(D)$ denoted $\mathcal{L}(Y)$. A monomial basis (for the vector space structure) can be computed as in Theorem 47 or in a generalised way as follows.

Consider the extended ring $\mathcal{K}[x_1, \ldots, x_d, y_1, \ldots, y_l]$. Let A be the matrix for the term-ordering τ above and let σ be a term-ordering over $\mathcal{K}[y_1, \ldots, y_l]$ corresponding to the matrix B. A term-ordering over the extended ring is defined by the block-diagonal matrix

$$\rho = \begin{bmatrix} A & 0 \\ 0 & B \end{bmatrix}$$

CONDITIONAL EXPECTATION

In particular, a lexicographic ordering operates among the two blocks of variables x and y, that is $x^\alpha \succ_\rho y^\beta$ for all α and β. It can be shown that the term-ordering ρ is an elimination term-ordering and thus by the elimination theory of Section 2.9, the Gröbner basis with respect to ρ of the ideal generated by

$$\begin{cases} g_1(x), \ldots, g_k(x) \\ y_j - \sum_{\alpha \in L} c_{\alpha,j} x^\alpha \qquad j = 1 \ldots l \end{cases}$$

is of the form

$$\begin{cases} h_1(y), \ldots, h_m(y) \\ f_1(x,y), \ldots, f_q(x,y) \end{cases}$$

Again, the h polynomials above form a Gröbner basis of the image support of the functions $Y = (Y_j)_{j=1,\ldots,l}$, $D^* = Y(D)$. Note that in $\mathrm{Est}(D^*)$ there may be less elements than in $\mathrm{Est}(D)$, because now we have y-terms and the total number has to stay the same. Let $\{y^\beta : \beta \in L^*\}$ be the monomial basis of $\mathrm{Ideal}(D^*)$ and L be the full list of exponents. Then the ring of functions, Y over D^* coincides with the vector space of such monomials.

With this characterisation in mind, we use the following definition of conditional expectation.

Definition 60 *The conditional expectation of the random variable $Z \in \mathcal{L}(D)$ is defined as the unique $\hat{Z} \in \mathcal{L}(Y)$ such that*

$$\text{for all } G \in \mathcal{L}(Y) \qquad \mathrm{E}_P(ZG) = \mathrm{E}_P\left(\hat{Z}G\right)$$

Clearly it is enough to check all $G = Y^\beta$, $\beta \in L^*$.

If P is strictly positive over D, the conditional expectation is given in terms of the base uniform distribution by

$$\hat{Z} = \mathrm{E}_P(Z|Y_1, \ldots, Y_l) = \frac{\mathrm{E}_0(ZP|Y_1, \ldots, Y_l)}{\mathrm{E}_0(P|Y_1, \ldots, Y_l)}$$

which follows simply by the relation $\mathrm{E}_P(U) = \mathrm{E}_0(UP)$ and substitution.

We concentrate on computing $\hat{Z} = \mathrm{E}_0(Z|Y_1, \ldots, Y_l)$. Using the linearity and idempotency of the conditional expectation operator, we write the defining condition as

$$\text{for all } \beta \in L, \alpha \in L^- \qquad \mathrm{E}_0\left(X^\alpha Y^\beta\right) = \mathrm{E}_0\left(\sum_{\gamma \in L^*} \hat{c}_{\gamma\alpha} Y^\gamma Y^\beta\right)$$

where L^- is the set of all indices in L with zero entries in y. Each of these equations is a regression equation: we are estimating each X^α as a linear function of the Y^β, $\beta \in L^*$ at the points of the design. As the model space is a ring, the conditional expectation coincides with the linear least-squares regression.

Example 71 [Continuation of Example 63] Let $l = 1$ and Y be the sum $X_1 + X_2 + X_3$. Then with respect to the term-ordering, ρ, defined by the matrix over the extended space $\mathbb{R}[x_1, x_2, x_3, y]$

	x_1	x_2	x_3	y
	1	1	1	0
	0	0	-1	0
	0	-1	0	0
	0	0	0	1

The support $D^\star$ is $\{4, 5, 6, 7, 8\}$ with Gröbner bases $(y-4)(y-5)(y-6)(y-7)(y-8)$. With respect to the term-ordering ρ the full set Est is

$1, \quad x_2, \quad x_3,$
$x_2^2, \quad x_2x_3, \quad x_3^2,$
$x_2y, \quad x_3y, \quad x_2^2y,$
$x_2x_3^2, \quad x_2x_3y,$
$x_2y^2, \quad x_3^2y, \quad x_3y^2, \quad x_2x_3y^2, \quad x_2y^3, \quad x_3^2y^2, x_3y^3,$
$x_2y^4, \quad x_3y^4$
$y, \quad y^2, \quad y^3, \quad y^4$

For each $X^\alpha \in \{1, X_2, X_3, X_2^2, X_2X_3, X_3^2\}$ it remains to determine a function $\hat{Z}_\alpha = c_{0,\alpha} + c_{1,\alpha}y + c_{2,\alpha}y^2 + c_{3,\alpha}y^3 + c_{4,\alpha}y^4$ such that

$$\mathrm{E}_0\left(X^\alpha Y^\beta\right) = \mathrm{E}_0\left(\hat{Z}_\alpha Y^\beta\right) \tag{5.18}$$

for all $Y^\beta \in \{1, Y, Y^2, Y^3, Y^4\}$. Using the values in Table 5.1 the six systems of linear formulae in (5.18) are easily solved. The equations below show the coefficients for $\hat{Z}_\alpha$

$$\mathrm{E}_0\left(X_2|Y\right) = -107.5026224 + 77.18138112Y - 19.94449301Y^2$$
$$+ 2.265005828Y^3 - .09484265734Y^4$$

$$\mathrm{E}_0\left(X_3|Y\right) = 106.2237762 - 75.49388112Y + 19.74256993Y^2$$
$$- 2.253787879Y^3 + .09484265734Y^4$$

$$\mathrm{E}_0\left(X_2^2|Y\right) = -443.8009907 + 315.5786713Y - 81.92416958Y^2$$
$$+ 9.372086247Y^3 - .3951777389Y^4$$

$$\mathrm{E}_0\left(X_2X_3|Y\right) = 17.23921911 - 11.60314685Y + 3.120556527Y^2$$
$$- .3569347319Y^3 + .01580710956Y^4$$

$$\mathrm{E}_0\left(X_3^2|Y\right) = 421.5617716 - 303.2543706Y + 79.99111305Y^2$$
$$- 9.217074592Y^3 + .3905885781Y^4$$

Table 5.1 Values of $\mathrm{E}_0\left(X^\alpha Y^\beta\right)$ for Example 63.

	1	x_2	x_3	x_2^2	$x_2 x_3$	x_3^2
1	1	2	2	14/3	47/12	14/3
y^1	6	25/2	25/2	30	51/2	30
y^2	75/2	81	81	199	699/4	391/2
y^3	243	552	1083/2	1419	2529/2	1272
y^4	3291/2	4053	3717	11238	38659/4	16211/2
y^5	11781					
y^6	179415/2					
y^7	722763					
y^8	12151011/2					

Example 72 [Continuation of Example 64] Consider again the expectation conditional to the sum $y = x_1 + x_2 + x_3$. The image support D^* is $\{3, 4, 5, 6, 7, 8, 9\}$ with Gröbner basis $(y-3)(y-4)(y-5)(y-6)(y-7)(y-8)(y-9)$. The set of estimable terms for the given support with the additional constraint given by the sum and with respect to the term-ordering ρ above is

$$\begin{array}{cccc} 1 & x_2 & x_3 & y \\ x_2 y & x_3^2 & x_3 y & y^2 \\ x_3 y^2 & x_3 y^3 & & \\ y^3 & y^4 & y^5 & y^6 \end{array}$$

Using the values in Table 5.2, we can compute the conditional expectations of the basic terms

$$\mathrm{E}_0\left(X_2|Y\right) = 2.617475423 + .003011138705Y + .001389073323Y^2$$
$$+ .0005342423415Y^3 - .01018547732Y^4$$
$$.002553328738Y^5 - .0001585075427Y^6$$

$$\mathrm{E}_0\left(X_2 X_3|Y\right) = 6.880923605 + .01120505649Y + .005101676179Y^2$$
$$+ .001936314720Y^3 - .04906279881Y^4$$
$$+ .01228985613Y^5 - .0007605881219Y^6$$

and by symmetry $\mathrm{E}_0\left(X_3|X_1 + X_2 + X_3\right) = \mathrm{E}_0\left(X_2|X_1 + X_2 + X_3\right)$.

5.8 Algebraic representation of exponentials

Let a support D be given and let $\{g_1, \ldots, g_k\}$ be a Gröbner basis of the ideal Ideal(D) in the ring $k[x]$. If $\mathrm{Est}_\tau(D) = \{x^\alpha : \alpha \in L\}$ is the corresponding list of estimable terms, we write $[x] = [x^\alpha]_{\alpha \in L}$ where L has been ordered

Table 5.2 *Values of* $E_0\left(X^\alpha Y^\beta\right)$ *for Example 64.*

	1	x_2	x_3	$x_2 x_3$
1	1	31/14	31/14	5
y^1	41/7	97/7	97/7	465/14
y^2	258/7	643/7	643/7	3233/14
y^3	1724/7	4474/7	4474/7	23361/14
y^4	12108/7	32425/7	32425/7	174425/14
y^5	88616/7	243202/7	243202/7	1339305/14
y^6	671148/7	1877653/7	1877653/7	1504679/2
y^7	5229944/7			
y^8	41733348/7			
y^9	339667736/7			
y^{10}	2810401788/7			
y^{11}	23573207864/7			
y^{12}	199984864788/7			

for example according to τ. Each function of exponential form

$$\left[\exp\left(\sum_{\alpha\in L}\psi_\alpha a^\alpha\right)\right]_{a\in D} = \exp\left(Z[\psi]\right) \qquad (5.19)$$

where $[\psi] = [\psi_\alpha]_{\alpha\in L} \in k^N$ and $x \in D$ can be written as a polynomial in the ring

$$k(e^{\psi_\alpha a^\alpha} : a \in D, \alpha \in L)[x]/\mathrm{Ideal}(D)$$

where $k(e^{\psi_\alpha a^\alpha} : a \in D, \alpha \in L)$ is the set of rational functions in $e^{\psi_\alpha a^\alpha}$ and with coefficients in k. Then, there exists a unique polynomial representation of the function (5.19) as in Equation (5.1)

$$\exp(\psi^t x) = \sum_{\alpha\in L} e_\alpha(\psi)x^\alpha \qquad (5.20)$$

with $e_\alpha(\psi) \in k(e^{\psi_\alpha a^\alpha} : a \in D, \alpha \in L)$.

Writing $[e(\psi)] = [e_\alpha(\psi)]_{\alpha\in L}$ from Equation (5.3), we have

$$[e(\psi)] = Z^{-1}\exp\left(Z[\psi]\right)$$

By the addition rule of the exponential in Equation (5.20) and the reduction rule of Theorem 43, we obtain

$$\exp\left(\psi_1^t x\right)\exp\left(\psi_2^t x\right) = \sum_{\gamma\in L}\left(\sum_{\alpha,\beta\in L} e_\alpha(\psi_1)e_\beta(\psi_2)r(\alpha+\beta,\gamma)\right)x^\gamma$$

and then
$$e_\alpha(\psi_1 + \psi_2) = \sum_{\beta,\gamma \in L} e_\beta(\psi_1) e_\gamma(\psi_2) r(\beta + \gamma, \alpha)$$

or, in vector form
$$[e(\psi_1 + \psi_2)] = \left[[e(\psi_1)]^t [r(\beta + \gamma, \alpha)]_{\beta,\gamma \in L} [e(\psi_2)] \right]_{\alpha \in L}$$

Let ∂_β be the partial derivative operator with respect to the parameter ψ_β. By applying ∂_β to Equation (5.20) we obtain

$$\sum_{\alpha \in L} \partial_\beta e_\alpha(\psi) x^\alpha = \sum_{\alpha \in L} e_\alpha(\psi) x^{\alpha + \beta}$$
$$= \sum_{\alpha \in L} e_\alpha(\psi) \sum_{\gamma \in L} r(\alpha + \beta, \gamma) x^\gamma \quad (5.21)$$
$$= \sum_{\alpha \in L} \left(\sum_{\gamma \in L} r(\beta + \gamma, \alpha) e_\gamma(\psi) \right) x^\alpha$$

so that
$$\partial_\beta e_\alpha(\psi) = \sum_{\gamma \in L} r(\beta + \gamma, \alpha) e_\gamma(\psi)$$

or, in vector notation
$$\nabla e(\psi) = [\partial_\beta e_\alpha(\psi)]_{\beta, \alpha \in L} = [e(\psi)]^t [R(\gamma)]_{\gamma \in L}$$

5.9 Exponential form of a probability

The interest in exponentials stems from the exponential form of strictly positive probabilities on the support D. All positive probabilities have the form
$$p(X; \psi) = \exp\left(\psi^t X\right) = \sum_{\alpha \in L} e_\alpha(\psi) X^\alpha \quad (5.22)$$

with the normalisation condition
$$\sum_{\alpha \in L} e_\alpha(\psi) E_0 \left(X^\alpha\right) = \sum_{\alpha \in L} e_\alpha(\psi) m_\alpha = 1 \quad (5.23)$$

This imposes a linear restriction on the e_α's.

In the usual setting of exponential models (see Chapter 6), this restriction is presented in the following way. Let $L_0 = L \setminus \{0\}$, $\psi = [\psi_\alpha]_{\alpha \in L_0}$ and $X = [X^\alpha]_{\alpha \in L_0}$ with slight abuse of notation. Then
$$\begin{cases} P_\psi = \exp\left(\psi^t X - K(\psi)\right) \\ \exp\left(K(\psi)\right) = \sum_{\alpha \in L_0} e_\alpha(\psi) m_\alpha \end{cases}$$

is the form of the full exponential model on D and $K(\psi) = -\psi_0$ is called the cumulant generating function. Note that the base distribution is given by $\psi_\alpha = 0$, $\alpha \in L_0$ and is uniform. In the terminology of exponential models P can be considered as an exponential extension of the uniform. In the rest of this section we shall cover the interplay between this representation and the closely related standard moment generating functions.

As we have seen in Sections 5.1 to 5.4, the computations of moments both with respect to the uniform probability and with respect to general probabilities on the support D is crucial. Here we consider generating functions to add new tools and insight to the computation of moments.

5.9.1 The geometric generating function

Let $s = (s_1, \ldots, s_d)$ be real variables and consider the generating function

$$H(s,x) = \frac{1}{\prod_{i=1}^{d}(1 - s_i x_i)} = \sum_{\beta \geq 0} s^\beta x^\beta \qquad (5.24)$$

where $\beta = (\beta_1, \ldots, \beta_d)$. For fixed s, $H(s,x)$ may be interpreted in the usual way as a random variable in $\mathcal{L}(D)$,

$$H(s,x) = \sum_{\alpha \in L} b_\alpha(s) x^\alpha, \qquad x \in D$$

Here the $b_\alpha(s)$ are rational forms in s which only depend on D and τ. That is $H(s,x)$ can be seen as a polynomial over D with rational polynomial coefficients in s. In this way we have "folded" the infinite power series into a finite polynomial in the particular x^α, $\alpha \in L$.

We can express this in matrix notation. Writing $[h(s)]_D = [H(s,a)]_{a \in D}$ and $[b(s)]_L = [b_\alpha(s)]_{\alpha \in L}$ we have

$$[h(s)]_D = Z[b(s)]_L$$

and

$$[b(s)]_L = Z^{-1}[h(s)]_D$$

It is instructive to proceed in a different algebraic manner. We have

$$\sum_{\alpha \in L} b_\alpha(s) x^\alpha \prod_{i=1}^{d}(1 - s_i x_i) = 1 \qquad (5.25)$$

Considering the monomials which appear in multiplying out the left-hand side of Equation (5.25) we have terms

$$x^\alpha \prod_{i \in K} x_i$$

for all different index sets K. We then obtain potentially all x^α, $\alpha \in L$

together with monomials just "above" the upper boundary of L. This is made clear by the following diagram

$$
\begin{array}{cccccc}
\circ & \circ & \circ & & & \\
\cdot & \cdot & \circ & \circ & \circ & \\
\cdot & \cdot & \cdot & \cdot & \circ & \\
\cdot & \cdot & \cdot & \cdot & \circ & \\
\end{array}
$$

Note that in the reduction form of Equation (5.25) we could use only border terms, but we prefer a more general approach using all terms in $L + L$.

By writing

$$\sum_{\alpha \in L} \tilde{b}_\alpha x^\alpha = \prod_{i=1}^{d}(1 - x_i)$$

we obtain

$$1 = \left(\sum_{\alpha \in L} b_\alpha(s) x^\alpha\right)\left(\sum_{\beta \in L} \tilde{b}_\beta s^\beta x^\beta\right) = \sum_{\gamma \in L} x^\gamma \sum_{\alpha,\beta \in L} r(\alpha + \beta, \gamma) b_\alpha(s) s^\beta \tilde{b}_\beta$$

and equivalently

$$\sum_{\alpha,\beta \in L} r(\alpha + \beta, \gamma) b_\alpha(s) s^\beta \tilde{b}_\beta = 0 \quad \gamma \in L_0$$

$$\sum_{\alpha,\beta \in L} r(\alpha + \beta, 0) b_\alpha(s) s^\beta \tilde{b}_\beta = 1$$

This gives a means to compute b_α's without using the design matrix Z.

By Equation (5.24)

$$H(s, x) = \sum_{\beta \geq 0} s^\beta X^\beta = \sum_{\alpha \in L} b_\alpha(s) X^\alpha$$

and taking expectation with respect to a probability P on D, we have

$$H(s) = \mathrm{E}_P\left(H(s, X)\right) = \sum_{\beta \geq 0} s^\beta \mathrm{E}_P\left(X^\beta\right) = \sum_{\alpha \in L} b_\alpha(s) \mathrm{E}_P\left(X^\alpha\right)$$

The left-hand side is a type of moment generating function and the formula carries over all information about how higher-order moments "fold" into lower-order moments. This is an instance of *moment aliasing*.

5.9.2 Univariate moment generating function

Let D be the support and let $Y = \sum_{\alpha \in L} c_\alpha X^\alpha$ be a generic random variable in $\mathcal{L}(D)$ represented with respect to $\mathrm{Est}_\tau(D)$. The uniform moment generating function of Y is defined as

$$M_Y(t) = \mathrm{E}_0\left(\exp(tY)\right)$$

We derive a differential-algebraic system for M_Y. We start with the Taylor series expansion of the exponential

$$\exp(tY) = \sum_{k=0}^{\infty} \frac{t^k Y^k}{k!} \tag{5.26}$$

and compute

$$Y^k = \left(\sum_{\alpha \in L} c_\alpha X^\alpha\right)^k = \sum_{\alpha \in L} c_\alpha(k) X^\alpha$$

The coefficients $c_\alpha(k)$, $\alpha \in L$, $k \in \mathbb{Z}_+$ satisfy a recurrence relation

$$\sum_{\alpha \in L} c_\alpha(k+1) X^\alpha = \left(\sum_{\alpha \in L} c_\alpha(k) X^\alpha\right)\left(\sum_{\alpha \in L} c_\alpha X^\alpha\right)$$

$$= \sum_{\alpha,\beta \in L} c_\alpha(k) c_\beta X^{\alpha+\beta}$$

$$= \sum_{\alpha,\beta \in L} c_\alpha(k) c_\beta \sum_{\gamma \in L} r(\alpha+\beta, \gamma) X^\gamma$$

$$= \sum_{\gamma \in L} X^\gamma \sum_{\alpha,\beta \in L} c_\alpha(k) c_\beta r(\alpha+\beta, \gamma)$$

where we have used the folding relations of Theorem 43. The recurrence relation linking the coefficients of Y_k is then

$$c_\gamma(k+1) = \sum_{\alpha \in L} c_\alpha(k) \sum_{\beta \in L} c_\beta r(\alpha+\beta, \gamma) \qquad \gamma \in L \text{ and } k \in \mathbb{Z}_+ \tag{5.27}$$

Note that Relation (5.27) applies also to $k = 0$ if we define

$$c_\alpha(0) = \begin{cases} 1 & \text{for } \alpha = 0 \\ 0 & \text{for } \alpha \neq 0 \end{cases}$$

We indicate this with $c_L(0) = 1_0$. In fact it gives

$$\sum_{\alpha \in L} c_\alpha(0) \sum_{\beta \in L} c_\beta r(\alpha+\beta, \gamma) = \sum_{\beta \in L} c_\beta r(\beta, \gamma) = c_\gamma$$

The linear recurrence relation (5.27) can be written in matrix form as

$$[c_L(k+1)] = A\,[c_L(k)]$$

where

$$A = \left[\sum_{\beta \in L} c_\beta r(\alpha+\beta, \gamma)\right]_{\gamma, \alpha \in L} = \sum_{\beta \in L} c_\beta R(\beta)$$

Also

$$[c_L(k)] = A^k\,[1_0]$$

We now use Equation 5.27 to derive an algebraic expression for the exponential in Equation 5.26. We have

$$\exp(tY) = \sum_{k=0}^{\infty} \frac{t^k}{k!} \sum_{\alpha \in L} c_\alpha(k) X^\alpha$$

and

$$M_Y(t) = \sum_{k=0}^{\infty} \frac{t^k}{k!} \sum_{\alpha \in L} c_\alpha(k) m_\alpha$$

$$= \sum_{\alpha \in L} m_\alpha \sum_{k=0}^{\infty} c_\alpha(k) \frac{t^k}{k!}$$

We define

$$\phi_\alpha(t) = \sum_{k=0}^{\infty} c_\alpha(k) \frac{t^k}{k!}$$

and show that the vector function $\phi_L(t) = [\phi_\alpha(t)]_{\alpha \in L}$ satisfies a linear differential equation. By derivation and using again the recurrence relation in (5.27)

$$\frac{d}{dt}\phi_\alpha(t) = \sum_{k=0}^{\infty} c_\alpha(k+1) \frac{t^k}{k!}$$

$$= \sum_{k=0}^{\infty} \left(\sum_{\beta \in L} c_\beta(k) A_{\alpha,\beta} \right) \frac{t^k}{k!}$$

$$= \sum_{\beta \in L} A_{\alpha,\beta} \left(\sum_{k=0}^{\infty} c_\beta(k) \frac{t^k}{k!} \right)$$

$$= \sum_{\beta \in L} A_{\alpha,\beta} \phi_\beta(t)$$

As $[\phi_L(0)] = [c_L(0)] = [e_1]$, the linear differential system sought is

$$\begin{cases} \dfrac{d}{dt}[\phi_L(t)] = A[\phi(t)] \\ [\phi_L(0)] = [1_0] \\ M_Y = \phi_L^t[m_L] \end{cases} \qquad (5.28)$$

If we write the solution of the linear differential system as an exponential matrix, we obtain

$$M_Y(t) = [m]_L^t \, e^{tA} \, [1_0] \qquad (5.29)$$

5.9.3 Moments of terms in the exponential model

Note that the above cases are for the moments of the raw X_i. In the exponential family case we will be interested in moments of the monomials given by L_0 itself. Thus, in a manner similar to the previous sections, we write

$$\tilde{H}(\psi, x) = \frac{1}{\prod_{\alpha \in L_0}(1 - \psi_\alpha x^\alpha)}$$

$$= \prod_{\alpha \in L_0} \sum_{\beta \geq 0} (\psi_\alpha)^\beta x^{\alpha+\beta}$$

$$= \sum_{\alpha \in L} \tilde{b}_\alpha(\psi) x^\alpha$$

and

$$\tilde{H}_P(\psi) = \sum_{\alpha \in L} \tilde{b}_\alpha(s) \mathrm{E}_P(X^\alpha)$$

The computation of the polynomial representation of the rational function is as in the previous case.

From Equation (5.20)

$$M(\psi, x) = \exp\left\{\sum_{\alpha \in L_0} \psi_\alpha x^\alpha\right\}$$

$$= \sum_{\alpha \in L} e_\alpha(\psi) x^\alpha$$

and the moment generating function is

$$M_P(\psi) = \sum_{\alpha \in L} e_\alpha(\psi) \mathrm{E}_P(X^\alpha)$$

CHAPTER 6

Statistical modeling

6.1 Introduction

This chapter is devoted to the statistical applications of the polynomial encoding of probabilities on a sample space D described as a zero-dimensional variety. We bring together the results on algebraic modeling of Chapter 2 with the treatment of probability theory of Chapter 5 to discuss statistical modeling and analysis.

It will appear that the polynomial algebraic description is well suited to describe operations on statistics (random variables) and on the algebra of parameters, especially in particular cases, such as the lattice case. We must underline the two applications, namely computation with random variables versus computations with parameters; the latter being of a quite different complexity. In fact, in the first case we work on ideals of points, while in the second case we have to deal with generic algebraic varieties. The computational load in the second case is much higher, and most of the general algorithms available at the moment of writing this book are unable to really deal with the symbolic solution of larger problems, a typical case being the symbolic explicit solution of the maximum likelihood equations.

The solution we suggest consists in adopting a hybrid approach: use the algebraic approach where it works very well, for example in the description of the structure of models based on conditional independence assumptions, and switch to a numerical approach when the symbolic solution is computationally infeasible, as in the solution of some maximum likelihood equations. Many research fields that use a computational commutative algebra approach meet the same problem nowadays, and there is an important ongoing research effort to develop efficient hybrid algorithms. As far as this book is concerned, we restrict ourselves to these generic remarks and will not discuss the matter further.

After a description of how all basic problems of modeling and estimation are encoded in a polynomial way, we make, in Section 6.2, a systematic review of how basic results on statistical modeling and estimation are translated in our framework.

We close the present Chapter (and the book), with a long, detailed example (see 6.9 below) on the treatment of a special graphical model (see Figure 6.1), which leads us into a brief glimpse of toric ideals. This example shows the evidence for our overall conclusion as discussed above. Namely, the algebraic polynomial theory is the most appropriate presentation and computational method where the theory of designs is concerned. It supplies interesting, albeit mainly conceptual, tools for the study of statistical models. The effectiveness of the algebraic methods in statistical models relies either on special cases, such that the lattice case treated here, or on a drastic improvement of the efficiency of computer algebra algorithms. In the final Section 6.10, we make a number of concise statements pertaining to the structure of maximum likelihood equations in the case of a lattice sample space.

In this chapter, D is a design with N points, τ a term-ordering and Est $= \{x^\alpha : \alpha \in L\}$ the corresponding saturated set of identifiable terms. With L' or M we indicate a subset of L, $L_0 = L \setminus \{0\}$, $M_0 = M \setminus \{0\}$ and $L'_0 \subseteq L \setminus \{0\}$.

6.2 Statistical models

Statistics is mainly concerned with models, submodels and the relations between them. As a general rule, according to the scope of this book, we consider models and submodels that can be specified by an algebraic variety, Θ with respect to some set of parameters with ideal denoted by Ideal (Θ). We will call these sort of models *algebraic statistical models*. Being an algebraic model depends on the parameters used in the probability description. In fact, some parameter transformations can destroy the algebraic character of the parameterisation. In particular, an algebraic model can be *linear* when the algebraic variety is an affine subspace of the space of parameters in the saturated model.

The most basic example consists in using as parameters the value of the density function itself, that is, we have a vector parameter $p = (p_i : i = 1 \ldots N)$ with

$$p_i = \mathrm{P}(a_i), \quad a_i \in D, \quad \sum_{i=1}^N p_i = 1$$

where a_i, $i = 1 \ldots N$ is a numbering of design points, and p is restricted to some algebraic variety Π described by an ideal in the ring $k[p]$. The ideal of this variety will be specified by giving equations in addition to the normalisation condition. Note that the positivity condition $p_i \geq 0$ has to be considered outside the algebraic framework.

Many types of models involving independence, conditional independence, fixed marginals and so on are algebraic submodels with respect to the point probability parameters. An example in a graphical model is given in Equation (6.22) below.

Because the statistical models are described as algebraic varieties, the problem of finding a minimal set of free parameters, that is, a proper parameterisation, could be discussed in the framework of the parametric (rational) representation of algebraic varieties, as discussed for example in Cox, Little and O'Shea (1997). We will not discuss this further, and restrict ourselves in the following to specific examples, namely the polynomial encoding and the exponential. As a reference for the basic estimation theory, we recommend Lehmann (1983) and (1986), and Kiefer (1987).

6.2.1 Polynomial form

Let us consider a statistical model whose density with respect to the uniform probability on the support D is given in polynomial form as

$$p(x;\theta) = \sum_{\alpha \in L} \theta_\alpha x^\alpha \qquad (6.1)$$

with θ belonging to a variety Θ of $\mathbb{R}^N$; we assume Θ to be the variety of a prime Ideal(Θ) of the ring $\mathbb{R}[\theta_\alpha : \alpha \in L]$. The ideal includes the equation defining the normalising constant in Equation (6.1), then the parameter θ_0 can be computed as

$$\theta_0 = 1 - \sum_{\alpha \in L_0} \theta_\alpha m_\alpha$$

with $m_\alpha = E_0(X^\alpha)$. We can rewrite the polynomial encoding as

$$p(x;\theta) = 1 + \sum_{\alpha \in L_0} \theta_\alpha (x^\alpha - m_\alpha)$$

where now $\theta = (\theta_\alpha : \alpha \in L_0)$ and the indeterminate θ_0 has been eliminated from Ideal(Θ).

The θ parameters are connected by linear relationships to the moment parameters μ,

$$\mu_\beta = E_\theta(X^\beta)$$
$$= \sum_{\alpha \in L} Q(\alpha, \beta) \theta_\alpha \qquad (6.2)$$

with $Q(\alpha, \beta) = E_0(X^\alpha X^\beta)$, and to the probability parameters p,

$$\mu_\beta = E_\theta(X^\beta)$$
$$= \sum_{i=1}^{N} X^\beta(a_i) p_i \qquad (6.3)$$

Note also the relation giving $p(a;\theta)$ for each $a \in D$ as a function of the θ parameters:

$$p(a) = \sum_{\alpha \in L} \theta_\alpha X^\alpha(a) \qquad (6.4)$$

We could write Equation (6.2) in vector form as

$$[\mu] = Q[\theta]$$

Equation (6.3) as

$$[\mu] = Z^t[p]$$

and Equation (6.4) as

$$[p] = Z[\theta]$$

where $Z = [X^\alpha(a)]_{a \in D, \alpha \in L}$ is the design matrix and $Q = \dfrac{1}{\#D} Z^t Z$.

Any algebraic model in any of the parameters $[p]$, $[\theta]$, $[\mu]$ is also an algebraic model in the other two parameters because of the linear dependencies above. In particular, a *linear model*, with respect to any of these linearly equivalent parameterisations, is defined as a model where the variety of the parameters Θ is an affine subspace of $\mathbb{R}^N$ and can be described by linear equations. Using the term introduced by Amari, see Amari (1985), such models are *mixture statistical models*, because, given two probabilities on the model, the segment connecting the values of the parameters at each of them runs over all mixtures of the two given probabilities.

The following two examples are devoted to the illustration of sampling and sufficiency reduction when the polynomial form is used. The treatment of such items as independence and sufficiency are particularly interesting in the polynomial encoding, where general algorithms are not available in the current literature. The usual reference to factorisation does not suggest a constructive algorithm.

Example 73 [Independent marginals] Let us consider the simplest possible model of a two-dimensional sampling distribution, that is, two independent Bernoulli variables. We denote the success probabilities by p_1, p_2. The generic joint probability in polynomial form is

$$p(x, y; \theta) = \theta_{00} + \theta_{10} x + \theta_{01} y + \theta_{11} xy$$

with marginals

$$p_1(x; \theta) = \left(\theta_{00} + \frac{1}{2} \theta_{01} \right) + \left(\theta_{10} + \frac{1}{2} \theta_{11} \right) x$$

$$p_2(y; \theta) = \left(\theta_{00} + \frac{1}{2} \theta_{10} \right) + \left(\theta_{01} + \frac{1}{2} \theta_{11} \right) y$$

The ideal of the model is obtained by the normalising equation and the four equations obtained by equating the coefficients in

$$p(x, y; \theta) = p_1(x; \theta) p_2(y; \theta)$$

namely,

$$\begin{cases} 4 = \theta_{00} + 2\theta_{10} + 2\theta_{01} + \theta_{11} \\ \theta_{00} = \left(\theta_{00} + \frac{1}{2}\theta_{01}\right)\left(\theta_{00} + \frac{1}{2}\theta_{10}\right) \\ \theta_{10} = \left(\theta_{00} + \frac{1}{2}\theta_{10}\right)\left(\theta_{10} + \frac{1}{2}\theta_{11}\right) \\ \theta_{01} = \left(\theta_{00} + \frac{1}{2}\theta_{01}\right)\left(\theta_{01} + \frac{1}{2}\theta_{11}\right) \\ \theta_{11} = \left(\theta_{10} + \frac{1}{2}\theta_{11}\right)\left(\theta_{01} + \frac{1}{2}\theta_{11}\right) \end{cases} \quad (6.5)$$

A proper parameterisation can be given in terms of the success probability as

$$\begin{cases} p_1 = p_1(1;\theta) = \left(\theta_{00} + \frac{1}{2}\theta_{01}\right) + \left(\theta_{10} + \frac{1}{2}\theta_{11}\right) \\ p_2 = p_2(1;\theta) = \left(\theta_{00} + \frac{1}{2}\theta_{10}\right) + \left(\theta_{01} + \frac{1}{2}\theta_{11}\right) \end{cases} \quad (6.6)$$

Solving Equations (6.5) and (6.6) by reduction to triangular form for the monomial ordering `plex` with initial ordering

$$\theta_{00} \succ \theta_{10} \succ \theta_{01} \succ \theta_{11} \succ p_1 \succ p_2$$

we obtain

$$\begin{cases} -4\theta_{00} - 2\theta_{10} - 2\theta_{01} - \theta_{11} + 4 = 0, \\ -\frac{1}{2}\theta_{10} + \frac{1}{2}\theta_{01} + p_1 - p_2 = 0, \\ \frac{1}{2}\theta_{01} + 1/4\theta_{11} - p_2 + 1 = 0, \\ -\theta_{11} + 4p_1p_2 - 4p_1 - 4p_2 + 4 = 0 \end{cases}$$

From this we could solve for the θ's and by substitution obtain the properly parameterised version of the density.

Let us see what happens in the case $\theta = p_1 = p_2$. If we sum the sample space ideal, the model ideal and the probability ideal, we obtain the ideal generated by the polynomials

$$\begin{cases} x^2 - x, \\ y^2 - y, \\ 4 - (4\theta_{00} + 2\theta_{10} + 2\theta_{01} + \theta_{11}), \\ \theta_{00} - \left(\theta_{00} + \frac{1}{2}\theta_{01}\right)\left(\theta_{00} + \frac{1}{2}\theta_{10}\right), \\ \theta_{10} - \left(\theta_{00} + \frac{1}{2}\theta_{10}\right)\left(\theta_{10} + \frac{1}{2}\theta_{11}\right), \\ \theta_{01} - \left(\theta_{00} + \frac{1}{2}\theta_{01}\right)\left(\theta_{01} + \frac{1}{2}\theta_{11}\right), \\ \theta_{11} - \left(\theta_{10} + \frac{1}{2}\theta_{11}\right)\left(\theta_{01} + \frac{1}{2}\theta_{11}\right), \\ p - (\theta_{00} + \theta_{10}x + \theta_{01}y + \theta_{11}xy) \end{cases}$$

A Gröbner basis is

$$\begin{cases} -4\theta_{00} - 2\theta_{10} - 2\theta_{01} - \theta_{11} + 4, \\ -\frac{1}{2}\theta_{10} + \frac{1}{2}\theta_{01}, \\ \frac{1}{2}\theta_{01} + 1/4\theta_{11} - \theta + 1, \\ \theta_{11} + 4\theta^2 - 8\theta + 4, \\ 4xy\theta^2 - 8xy\theta + 4xy - 2x\theta^2 + 6x\theta - \\ 4x - 2y\theta^2 + 6y\theta - 4y + \theta^2 - \theta - p + 4 \end{cases}$$

The last polynomial is the representation of the probability density. In the corresponding equation, the part equal to p factors out as follows:

$$p = (2x\theta - 2x - \theta + 2)(2y\theta - 2y - \theta + 2)$$

If we order with respect to the powers of θ, we obtain the representation

$$\begin{aligned} p = &(4xy - 4x - 4y + 4) + \\ &(-8xy + 6x + 6y - 1)\theta + \\ &(+4xy - 2x - 2y + 1)\theta^2 \end{aligned}$$

where the polynomial coefficients are a set of sufficient statistics. Actually an elementary analysis shows that $T(x,y) = x + y$ is a sufficient statistic because on D we have $(x+y)^2 = (x+y) + 2xy$, that is $xy = \frac{1}{2}T(x,y)^2 - \frac{1}{2}T(x,y)$. The generalisation of this argument, leading to a general algorithm for sufficiency, is described in the next section.

Example 74 [Independence on a single cell] Let us consider a 2×3 table, with sample space $\{1,2\} \times \{1,2,3\}$. Let us consider a model where inde-

pendence is assumed for the cell $(1,1)$. In terms of the p parameters the model is described by the equations
$$\begin{cases} p(1,1) + p(2,1) + p(1,2) + p(2,2) + p(1,3) + p(2,3) = 1 \\ (p(1,1) + p(1,2) + p(1,3))(p(1,1) + p(2,1)) = p(1,1) \end{cases}$$
The sample space is described by the Gröbner basis
$$\begin{cases} (x-1)(x-2), \\ (y-1)(y-2)(y-3) \end{cases}$$
and the list of estimable terms is
$$1, \quad x, \quad y, \quad y^2, \quad xy, \quad xy^2$$
The design matrix is

	1	x	y	y^2	xy	xy^2
11	1	1	1	1	1	1
21	1	2	1	1	2	2
12	1	1	2	4	2	4
22	1	2	2	4	4	8
13	1	1	3	9	3	9
23	1	2	3	9	6	18
Total	6	9	12	28	18	42

Using the relation $[p] = Z[\theta]$ we find, using dot notation for sum,
$$\begin{cases} p(1,\cdot) = 7\theta_{11} + 7\theta_{21} + 18\theta_{12} + 50\theta_{22} + 18\theta_{13} + 50\theta_{23} \\ p(\cdot,1) = 3\theta_{11} + 5\theta_{21} + 3\theta_{12} + 3\theta_{22} + 5\theta_{13} + 5\theta_{23} \\ p(1,1) = \theta_{11} + \theta_{21} + \theta_{12} + \theta_{22} + \theta_{13} + \theta_{23} \end{cases}$$

Collecting all the terms together, the model is defined by the ideal generated by
$$\begin{cases} x^2 - 3x + 2, \\ y^3 - 6y^2 + 11y - 6, \\ 6\theta_{11} + 9\theta_{21} + 12\theta_{12} + 28\theta_{22} + 18\theta_{13} + 42\theta_{23} - 6, \\ \theta_{11} + \theta_{21} + \theta_{12} + \theta_{22} + \theta_{13} + \theta_{23} - \\ \quad [(7\theta_{11} + 7\theta_{21} + 18\theta_{12} + 50\theta_{22} + 18\theta_{13} + 50\theta_{23}) \\ \quad (3\theta_{11} + 5\theta_{21} + 3\theta_{12} + 3\theta_{22} + 5\theta_{13} + 5\theta_{23})] \end{cases}$$

In this system of equations, the sample variables x and y are separated from the θ variables. In particular, no reduction is possible, except the expression of θ_{11} as a function of the remaining θ's by the normalising equation. But if we introduce the new indeterminate p, representing the values of the probability (but distinct from the probability parameter), by adding the equation

$$p = \theta_{11} + \theta_{21}x + \theta_{12}y + \theta_{22}y^2 + \theta_{13}xy + \theta_{23}xy^2$$

we can obtain in the Gröbner basis polynomials in x, y and θ. See below for the computation of the sufficient statistics.

6.2.2 Sufficiency in polynomial models

We recall the basic theory of sufficiency in a form suitable for our purpose. Let p_θ, $\theta \in \Theta$, be a statistical model such that there exists a reference value of the parameter θ_0 for which all the likelihoods

$$\frac{p_\theta}{p_{\theta_0}} = \ell_\theta, \quad \theta \in \Theta$$

are defined. More precisely, the likelihood is a random variable with values into the parametric functions,

$$D \ni x \mapsto (\theta \mapsto \ell_\theta(x))$$

The σ-algebra generated by this random variable, that is to say the σ-algebra generated by all the random variables ℓ_θ, $\theta \in \Theta$, is the *sufficient σ-algebra*. The space of bounded random variables measurable on this σ-algebra is a ring with unity, the *sufficient ring*. Any set of random variables $\{T\}$ which generates the sufficient σ-algebra on the sufficient ring is a set of *sufficient statistics*.

In our case, the sample space D is finite, and the sufficient ring is a subring of the ring of random variables $\mathcal{L}(D)$. Let us assume that the model is algebraic, and consider the quotient ring

$$\frac{\mathbb{R}[\theta_\alpha : \alpha \in L_0]}{\text{Ideal}(\Theta)}$$

We call this the ring of *algebraic parametric functions*. As a vector space, and given a monomial ordering, this ring has a monomial basis $\{\theta^\beta : \beta \in M\}$. Every parametric function is characterised by the list of θ^β's coefficients. In other words, we have a different representation of the likelihood as

$$T_\beta : D \ni x \mapsto T_\beta(x), \quad \beta \in M$$

where

$$\ell_\theta(x) = \sum_{\beta \in M} T_\beta(x) \theta^\beta$$

The set $\{T_\beta : \beta \in M\}$ contains sufficient statistics for which there is a computational algorithm.

The given sufficient statistics, as the following example shows, is in most cases further reducible, by considering a subset of the T_β's generating the

STATISTICAL MODELS

same ring. If we consider the polynomial mapping (sufficiency reduction)

$$[T_\beta]_{\beta \in M} : D \ni x \mapsto [t_\beta]_{\beta \in M} = [T_\beta(x)]_{\beta \in M} \in \mathbb{R}^M$$

this mapping induces a partition on D into sets where the likelihood is constant. By computing the image design $D^\star$, it is possible to find a monomial basis for the indeterminates T_β's, possibly obtaining a further reduction of the number of functions in the sufficient statistics. The following example illustrates the procedure.

Example 75 [Continuation of Example 73] The model variety Θ is defined by Equations (6.5). The sum of the model ideal and the support ideal is generated by the polynomials

$$\begin{cases} 4 - \theta_{00} + 2\theta_{10} + 2\theta_{01} + \theta_{11}, \\ \theta_{00} - \left(\theta_{00} + \frac{1}{2}\theta_{01}\right)\left(\theta_{00} + \frac{1}{2}\theta_{10}\right), \\ \theta_{10} - \left(\theta_{00} + \frac{1}{2}\theta_{10}\right)\left(\theta_{10} + \frac{1}{2}\theta_{11}\right), \\ \theta_{01} - \left(\theta_{00} + \frac{1}{2}\theta_{01}\right)\left(\theta_{01} + \frac{1}{2}\theta_{11}\right), \\ \theta_{11} - \left(\theta_{10} + \frac{1}{2}\theta_{11}\right)\left(\theta_{01} + \frac{1}{2}\theta_{11}\right), \\ x^2 - x, \\ y^2 - y \end{cases}$$

As we expect, the leading terms of a Gröbner basis for the monomial order tdeg $(\theta_{00} \succ \theta_{11} \succ \theta_{10} \succ \theta_{01} \succ x \succ y)$ (chosen for symmetry) are θ_{00}, θ_{11}, θ_{10}^2, x^2, y^2, so that the monomial basis involves θ_{10}, θ_{01}, x, y, xy only. The normal form of the probability density, ordered with respect to the free parameters θ_{10} and θ_{01}, is

$$p(x, y; \theta) = \frac{4}{5}(1 + xy)$$
$$+ \left[x - \frac{2}{5}(1 + xy)\right]\theta_{10}$$
$$+ \left[y - \frac{2}{5}(1 + xy)\right]\theta_{01}$$

In this case there is no reduction for sufficiency, because the ring generated by $x, y, 1 + xy$ and the constants is the full quotient ring $\mathcal{L}(D)$.

But assume we add the equation implying the equality of marginals, $\theta_{10} = \theta_{01} = \theta$. Then, the Gröbner basis has leading terms θ_{00}, θ_{10}, θ_{01}, θ_{11}, θ^2, x^2, y^2. The monomial basis involves only the parameter θ. The normal form of the probability, ordered according to θ is

$$p(x,y;\theta) = \frac{4}{5}(1+xy)$$
$$+ \left[x + y - \frac{4}{5}(1+xy)\right]\theta$$

In this case the ring generated is strictly a subring of the quotient ring, and $x + y$ is a sufficient statistic because $xy = \frac{1}{2}(x+y)^2 - 1$, on D.

6.2.3 Exponential form

In the exponential parameterisation, we consider mainly linear models of the form

$$p(x;\psi) = \exp\left(\sum_{\alpha \in L'_0} \psi_\alpha x^\alpha - K(\psi)\right)$$

where L' is an order ideal of the list of estimable terms. This is not a restriction, because it is always possible to make a change of indeterminate, assigning a new indeterminate to each polynomial in the model. In fact, assume we are interested in the linear model in exponential form

$$p(x;\psi) = \exp\left(\sum_{j=1}^{d} \psi_j F_j - K(\psi)\right)$$

where F_j, $j = 1, \ldots, d$ are linearly independent functions on D. If we introduce the new indeterminates y_j, $j = 1, \ldots, d$, and add to the sample space ideal the polynomials $y_j - F_j$, $j = 1, \ldots, d$, we reduce to the standard situation. In specific cases, other reductions to this case are possible, using a smaller number of new indeterminates.

6.3 Generating functions and exponential submodels

In this section we display the algebraic computations related to the moment generating functions of sufficient statistics of an exponential model. We refer to general treatises such as Barndorff-Nielsen and Cox (1989) and (1994) for the role of this key quantity in estimation theory.

The moment generating function of the X^α, $\alpha \in L'_0$, under the exponential model

$$P_\psi = \exp\left(\sum_{\alpha \in L'_0} \psi_\alpha X^\alpha - K(\psi)\right)$$

where $K(\psi) = K(\psi_\alpha : \alpha \in L'_0) = -\psi_0$, is

$$\begin{aligned}
M_\psi(s) &= \mathrm{E}_\psi\left(\exp\left(\sum_{\alpha \in L'_0} s_\alpha X^\alpha\right)\right) \\
&= \mathrm{E}_0\left(\exp\left(\sum_{\alpha \in L'_0} (s_\alpha + \psi_\alpha)X^\alpha - K(\psi)\right)\right) \\
&= \exp\left(K(s+\psi) - K(\psi)\right) \\
&= \frac{M_0(s+\psi)}{M_0(\psi)}
\end{aligned}$$

Now we use again the polynomial form of the exponential as a linear combination of X^α, $\alpha \in L$, namely

$$\exp\left(\sum_{\alpha \in L'_0} \psi_\alpha x^\alpha\right) = \sum_{\beta \in L} e_\beta(\psi) x^\beta$$

and the representation of expectations using raw moments

$$M_0(\psi) = \sum_{\beta \in L} e_\beta(\psi) m_\beta$$

so that

$$M_\psi(s) = \frac{\sum_{\beta \in L} e_\beta(s+\psi) m_\beta}{\sum_{\beta \in L} e_\beta(\psi) m_\beta} \tag{6.7}$$

Now at ψ, that is, with respect to P_ψ, the following identity holds

$$\mathrm{E}_\psi\left(X^\beta\right) = \frac{\partial_\beta M_0(\psi)}{M_0(\psi)} = \partial_\beta K(\psi)$$

where ∂_β means partial derivative with respect to ψ_β. From Equation (5.3) we have

$$\partial_\beta M_0(\psi) = \mathrm{E}_0\left(\partial_\beta e^{\sum_{\alpha \in L'} \psi_\alpha X^\alpha}\right) = \mathrm{E}_0\left(\partial_\beta \sum_{\alpha \in L} e_\alpha(\psi) X^\alpha\right)$$

We can proceed in two equivalent ways: differentiating the $e_\alpha(\psi)$ or substituting directly the value of the derivatives of X^β with respect to ψ_β, $\beta \in L'_0$.

We obtain, respectively

$$\partial_\beta M_0(\psi) = \mathrm{E}_0\left(\sum_{\alpha \in L} \partial_\beta e_\alpha(\psi) X^\alpha\right) = \sum_{\alpha \in L} \partial_\beta e_\alpha(\psi) m_\alpha$$

and

$$\partial_\beta M_0(\psi) = \mathrm{E}_0\left(X^\beta \sum_{\alpha \in L} e_\alpha(\psi) X^\alpha\right)$$
$$= \mathrm{E}_0\left(\sum_{\alpha \in L} e_\alpha(\psi) X^{\alpha+\beta}\right)$$
$$= \sum_{\alpha \in L} e_\alpha(\psi) Q(\alpha, \beta)$$

This implies

$$\mathrm{E}_\psi\left(X^\beta\right) = (M_0(\psi))^{-1} \sum_{\gamma \in L} \left(\sum_{\alpha \in L} e_\alpha(\psi) r(\alpha + \beta, \gamma)\right) m_\gamma$$

Now computing directly in

$$M_\psi(s) = \mathrm{E}_\psi\left(\exp\left(\sum_{\alpha \in L'} s^\alpha X_\alpha\right)\right)$$
$$= \mathrm{E}_\psi\left(\sum_{\beta \in L} e_\beta(s) X_\beta\right)$$
$$= \sum_{\beta \in L} e_\beta(s) \mathrm{E}_\psi\left(X^\beta\right)$$

and by substitution for $\mathrm{E}_\psi\left(X^\beta\right)$

$$M_\psi(s) = M_0(\psi)^{-1} \sum_{\beta \in L} e_\beta(\psi) \sum_{\gamma \in L} \left(\sum_{\alpha \in L} e_\alpha(\psi) r(\alpha + \beta, \gamma) m_\gamma\right)$$
$$= M_0(\psi)^{-1} \sum_{\gamma \in L'} e_\gamma(s + \psi) m_\gamma \quad \text{[from (6.7)]}$$

we have obtained two different representations. But equating coefficients of m_γ we have

$$e_\gamma(s + \psi) = \sum_{\beta \in L}\sum_{\alpha \in L} e_\beta(s) e_\alpha(\psi) r(\alpha + \beta, \gamma)$$

which returns also

$$\partial_\beta e_\gamma(\psi) = \sum_{\alpha \in L} e_\alpha(\psi) r(\alpha + \beta, \gamma)$$

These are, respectively, functional and differential equations coming from the original interpolator (5.3).

Note that they are distribution free in the sense that $r(\alpha + \beta, \gamma)$ only depends on D and τ.

6.4 Likelihoods and sufficient statistics

Consider the exponential model with distribution given by

$$P_\psi = \exp\left(\sum_{\alpha \in L_0'} \psi_\alpha X^\alpha - K(\psi)\right) \tag{6.8}$$

where $K(\psi)$ is the cumulant generating function of the X^α, $\alpha \in L_0'$ is a subset of L_0 as defined in Section 5.8, and we have taken the base distribution to be uniform. In statistical estimation we take an independent identically distributed sample s of size N. Thus we consider a sample space $S = D^N$.

Note that statements like $a \in s$ should be carefully distinguished from $a \in D$, which expresses membership of the support.

The likelihood for s in the exponential submodels in Equation (6.8) is

$$L(\psi, s) = \prod_{a \in s} \exp\left(\sum_{\alpha \in L_0'} \psi_\alpha a^\alpha - K(\psi)\right)$$

$$= \exp\left(\sum_{\alpha \in L_0'} \psi_\alpha \sum_{a \in s} \psi_\alpha a^\alpha - NK(\psi)\right)$$

The log-likelihood is

$$\log L(\psi, s) = \ell(\psi, s) = \sum_{\alpha \in L_0'} \psi_\alpha \sum_{a \in s} a^\alpha - NK(\psi)$$

According to standard theory, the quantities

$$T_\alpha = \sum_{a \in s} a^\alpha, \quad \alpha \in L_0'$$

are the sufficient statistics. The maximum likelihood equations are

$$\sum_{a \in s} a^\alpha = N \frac{\partial K}{\partial \psi_\alpha}(\psi)$$

$$= N \mathrm{E}_\psi(X^\alpha)$$

$$= N\mu_\alpha, \quad \alpha \in L_0'$$

or

$$\hat{\mu}_\alpha = \bar{\mu}_\alpha$$

$$= \frac{1}{N} \sum_{a \in s} a^\alpha$$

$$= \frac{1}{N} T_\alpha \quad \alpha \in L_0'$$

where $\bar{\mu}_\alpha$ is the notation for the sample mean of X^α and $\hat{\mu}_\alpha$ is the notation for the estimated value.

In vector notation we write

$$[\hat{\mu}_\alpha]_{\alpha \in L_0'} = [\bar{\mu}_\alpha]_{\alpha \in L_0'}$$

The parameters $[\psi]$ are free to assume values over the reals. The model induces restrictions on the parameters $[\theta]$ and to the individual probabilities $[p]$, leading to unique solutions to

$$[\bar{\mu}] = Q_{L',L}[\hat{\theta}] = \frac{1}{\#D} Z_1^t [\hat{p}]$$

where

$$Q_{L',L} = \left[E_0 \left(X^{\alpha+\beta} \right) \right]_{\alpha \in L'; \beta \in L}$$

and $Z_1 = [X^\alpha(a)]_{a \in D; \alpha \in L'}$ is the design matrix of the model. The restrictions on the p's are

$$[\hat{p}] = \left[\exp \left(\sum_{\alpha \in L_0'} \hat{\psi}_\alpha x^\alpha - K(\hat{\psi}) \right) \right]_{x \in D}$$
$$= \exp \left(Z_1[\hat{\psi}] \right)$$

where a real function of a vector is computed componentwise. Thus for the vector $[\hat{\psi}]$ of $\hat{\psi}_\alpha$, $\alpha \in L_0$,

$$Z_1^t Z_1 [\hat{\psi}] = Z_1^t \log[\hat{p}]$$

and

$$[\hat{\psi}] = \left(Z_1^t Z_1 \right)^{-1} Z_1^t \log \left(Z_1[\hat{p}] \right)$$

Let ∂_β be the partial differential operator corresponding to the index β (as defined in Section 5.8). We can develop an analysis based on the interpolation methods used in Section 5.9.3. The expectation of X^β at $[\psi]$, that is under the distribution in Section 6.3, is

$$\mu_\beta = E_\psi \left(X^\beta \right) = \frac{\partial_\beta M_0(\psi)}{M_0(\psi)} = \partial_\beta K(\psi) = K_{(\beta),\psi}$$

where $K_{(\beta),\psi}$ is the first-order cumulants of X^β at $[\psi]$ and $M_0(\psi)$ is the moment generating function under the distribution in Equation (6.8). Now consider the interpolator

$$\exp \left(\sum_{\alpha \in L_0'} \psi_\alpha x^\alpha \right) = \sum_{\alpha \in L} e_\alpha(\psi) x^\alpha$$

By the uniqueness of the interpolation, the $e_\alpha(\psi)$ are unique. Thus,

$$\partial_\beta \exp\left(\sum_{\alpha \in L_0'} \psi_\alpha x^\alpha\right) = \partial_\beta \sum_{\alpha \in L} e_\alpha(\psi) x^\alpha$$

$$= \sum_{\alpha \in L} \partial_\beta e_\alpha(\psi) x^\alpha$$

By direct differentiation

$$\partial_\beta \exp\left(\sum_{\alpha \in L} \psi_\alpha x^\alpha\right) = x^\beta \exp\left(\sum_{\alpha \in L} \psi_\alpha x^\alpha\right)$$

$$= \sum_{\alpha \in L} e_\alpha(\psi) x^{\alpha+\beta}$$

Now $x^{\alpha+\beta}$ can itself be uniquely interpolated:

$$x^{\alpha+\beta} = \sum_{\gamma \in L} r(\alpha+\beta, \gamma) x^\gamma$$

This yields

$$\sum_{\alpha \in L} \partial_\beta e_\alpha(\psi) x^\alpha = \sum_{\alpha \in L} e_\alpha(\psi) \sum_{\gamma \in L} r(\alpha+\beta, \gamma) x^\gamma$$

Equating coefficients

$$\partial_\beta e_\alpha(\psi) = \sum_{\gamma \in L} e_\gamma(\psi) r(\gamma+\beta, \alpha)$$

from which we deduce

$$\partial_\beta M_0(\psi) = \partial_\beta E_0\left(\exp\left(\sum_{\alpha \in L_0'} \psi_\alpha X^\alpha\right)\right)$$

$$= E_0\left(\partial_\beta \exp\left(\sum_{\alpha \in L_0'} \psi_\alpha X^\alpha\right)\right)$$

$$= \sum_{\alpha \in L} \partial_\beta e_\alpha(\psi) m_\alpha$$

$$= \sum_{\alpha \in L} \sum_{\gamma \in L} e_\gamma(\psi) r(\gamma+\beta, \alpha) m_\alpha$$

$$= \sum_{\gamma \in L} e_\gamma(\psi) \sum_{\alpha \in L} r(\gamma+\beta, \alpha) m_\alpha$$

Then, the likelihood equation becomes

$$\bar{\mu}_\beta = \frac{\partial_\beta M_0(\hat{\psi})}{M_0(\hat{\psi})} \tag{6.9}$$

or, noting that $M_0(\psi) = \sum_{\alpha \in L} e_\alpha(\psi) m_\alpha$

$$\left(\sum_{\alpha \in L} e_\alpha(\hat{\psi}) m_\alpha\right) \bar{\mu}_\beta = \sum_{\gamma \in L} e_\gamma(\hat{\psi}) \left(\sum_{\alpha \in L} r(\gamma + \beta, \alpha) m_\alpha\right) \quad \beta \in L' \quad (6.10)$$

Note that $e_{(0,\ldots,0)}(\psi) = 1$. These equations are potentially solvable for ψ_α, $\alpha \in L_0$. However, without further analysis they remain in non-algebraic form because the vector of $e_\alpha(\psi)$, $\alpha \in L$ has $\#D$ components: it cannot be evaluated by linear operations alone because $\#L' < \#D = \#L$. The additional restriction stems from the restrictions on ψ_α, $\alpha \in L_0$.

If we compare this development to Section 6.3 and the matrix development at the beginning of this section we have

$$\theta_\alpha = e_\alpha(\psi) \left(M_0(\psi)\right)^{-1} \quad \alpha \in L \quad (6.11)$$

and

$$[\hat{\mu}] = Q[\hat{\theta}]$$
$$= Q[e(\hat{\psi})] \left(\tilde{M}_0(\hat{\psi})\right)^{-1}$$

where $[e(\hat{\psi})]$ is the vector of $e_\alpha(\hat{\psi})$, $\alpha \in L_0$, giving

$$M_0(\hat{\psi})[\bar{\mu}] = Q[e(\hat{\psi})] \quad (6.12)$$

which is the matrix form of Equation (6.9).

The three equivalent representations of Equations (6.9), (6.10), (6.12) do not necessarily yield closed form solutions for $\hat{\psi}_\alpha$, $\alpha \in L'_0$, despite the fact that the $\hat{\psi}_\alpha$ can be computed in terms of the $\hat{\theta}_\alpha$. However, Level 4 algebrisation mentioned in Section 5.8 can assist.

Thus let $\zeta_\alpha = \exp(\psi_\alpha)$, $\alpha \in L_0$. Then, the normal equations can be written

$$\sum_{x \in D} x^\beta \prod_{\alpha \in L_0} \hat{\zeta}_\alpha^x = \bar{x}^\beta \quad \beta \in L'_0$$

$$\sum_{x \in D} \hat{\zeta}_\alpha^x = 1$$

Example 76 Consider $D = \{0, 1, 2\}$ and the model

$$P_\psi = \exp(X\psi_1 - K(\psi_1))$$

Then, $\psi_0 = -K(\psi_1)$, $\zeta_0 = \exp(-K(\psi_1))$ and $\zeta_1 = \exp(\psi_1)$. The equations are

$$\zeta_0 \left(\zeta_1 + 2\zeta_1^2\right) = \bar{\mu}_1$$
$$\zeta_0 \left(1 + \zeta_1 + \zeta_1^2\right) = 1$$

which are easily solved for (ζ_0, ζ_1) and hence for ψ_0 and ψ_1.

6.5 Score function and information

In this section we discuss some of the most basic quantities involved in statistical inference connected to the exponential model and submodels.

The score function is the gradient of the log-likelihood: $\ell(\psi_0, S)$ of Section 6.4

$$U_\alpha^{S;\psi} = \frac{\partial \ell(S)}{\partial \psi_\alpha} = \sum_{a \in s} X^\alpha(a) - N \frac{\partial K(\psi)}{\partial \psi_\alpha} \qquad \alpha \in L_0'$$

and in vector notation we write

$$U^{S;\psi} = \left[\frac{\partial \ell}{\partial \psi_\alpha} \right]_{\alpha \in L_0'}$$

The information matrix is (minus) the Hessian of the log-likelihood:

$$N \left[\frac{\partial^2 K(\psi)}{\partial \psi_\alpha \partial \psi_\beta} \right]_{\alpha, \beta \in L_0'} = \mathrm{Cov}_\psi\left(U^{S;\psi}\right)$$

In this case this is also the Fisher, or expected, information since, it does not depend on X.

Note that from Section 5.8 we have

$$\mathrm{Cov}_\psi\left(U^{S;\psi}\right) = N \left[\mathrm{Cov}_\psi\left(X^\alpha, X^\beta\right)\right]_{\alpha,\beta \in L_0'}$$

$$= N \left(\sum_{\gamma \in L} r(\alpha+\beta, \gamma) \mu_\gamma - \mu_\alpha \mu_\beta \right)$$

$\alpha, \beta \in L_0$.

The "estimated" information is obtained by evaluation at the maximum likelihood estimator $\hat{\psi}$ giving

$$N \left(\sum_{\gamma \in L} r(\alpha+\beta, \gamma) \hat{\mu}_\gamma - \hat{\mu}_\alpha \hat{\mu}_\beta \right)$$

Observe that $\hat{\mu}_\alpha = \bar{\mu}_\alpha$ only if $\alpha \in L'$. In matrix notation, using Equation (6.11)

$$[\hat{\mu}_\gamma]_{\gamma \in L} = Z^t[\hat{p}]$$
$$= Q[\hat{\theta}]$$
$$= Q[e(\hat{\psi})] M_0(\hat{\psi})$$

It is straightforward to derive the score functions for other parameterisations such as θ or ζ.

6.6 Estimation: lattice case

We review the implications of the lattice case for estimation. Let us consider a sample space D where each sample point is integer valued. Note that in this case
$$\text{Est}_\tau(D) = \{x^\alpha : \alpha \in L\}$$
contains integer-valued functions.

Let us consider the exponential model associated with the terms of a sublist $L' \subset L$ containing 0 and the corresponding powers X^α, $\alpha \in L'_0$. The design matrix Z has a sub-matrix Z_1 with columns restricted to L'.

If we introduce the new parameters (as in Section 6.4)
$$\zeta_\beta = \exp(\psi_\beta), \qquad \alpha \in L'$$
probabilities will be expressed as
$$p(a) = \zeta_0 \prod_{\beta \in L'} \zeta_\beta^{X^\beta(a)}, \qquad a \in D$$

Note that the exponents of the ζ monomials are the rows of the matrix Z_1. In a "parallel" notation we could write $[p] = \zeta^{Z_1}$. If we proceed to the computation of the moments of the sufficient statistics as a function of the ζ parameters, we obtain for all $\beta \in L'$
$$\mu_\beta = \sum_{a \in D} X^\beta(a) p(a)$$
$$= \sum_{a \in D} X^\beta(a) \zeta_0 \prod_{\beta \in L'} \zeta_\beta^{X^\beta(a)}$$

Using again the array notation as above, we could write
$$[\mu] = Z_1^t [\zeta^{Z_1}]$$

Equality of sample moments $\bar{\mu}$ and estimated moments required for the maximum likelihood estimation gives the following set of normal equations
$$\bar{\mu}_\beta = \sum_{a \in D} X^\beta(a) p(a)$$
$$= \sum_{a \in D} X^\beta(a) \zeta_0 \prod_{\beta \in L'} \zeta_\beta^{X^\beta(a)}$$

for $\beta \in L'$. It is a set of algebraic equations, whose positive solutions are unique if the sample moments have values belonging to the exponential model, which happens in the absence of zero values.

Example 77 The list of terms $\{1, x_1, x_2, x_1 x_2\}$, that is
$$L = \{(0,0),\ (1,0),\ (0,1),\ (1,1)\}$$
gives rise to the saturated exponential model for the binary design $D =$

ESTIMATION: LATTICE CASE

$\{0,1\}^2$ for any term-ordering given by

$$p(x_1, x_2; \psi_{10}, \psi_{01}, \psi_{11}) = \exp\left(\psi_{10} x_1 + \psi_{01} x_2 + \psi_{11} x_1 x_2 - K(\psi_{10}, \psi_{01}, \psi_{11})\right)$$

Consider the model associated with the subset $\{x_2, x_1 x_2\}$, giving

$$L' = \{(0,1),\ (1,1)\}$$

that is the exponential model

$$p(x_1, x_2; \psi_{01}, \psi_{11}) = \exp\left(\psi_{01} x_2 + \psi_{11} x_1 x_2 - K(\psi_{01}, \psi_{11})\right)$$

Consider independent sampling.
The sufficient statistics are

$$T_1 = x_2,\ T_2 = x_1 x_2$$

The matrices Z and Z_1 for the sample points ordered as

$$00 \prec 10 \prec 01 \prec 11$$

become

$$Z = \begin{bmatrix} 1 & 0 & 0 & 0 \\ 1 & 1 & 0 & 0 \\ 1 & 0 & 1 & 0 \\ 1 & 1 & 1 & 1 \end{bmatrix},\quad Z_1 = \begin{bmatrix} 1 & 0 & 0 \\ 1 & 0 & 0 \\ 1 & 1 & 0 \\ 1 & 1 & 1 \end{bmatrix}$$

We introduce the ζ parameters

$$\zeta_{00} = \exp\left(-K(\psi_{01}, \psi_{11})\right)$$
$$\zeta_{01} = \exp\left(\psi_{01}\right)$$
$$\zeta_{11} = \exp\left(\psi_{11}\right)$$

and obtain the following expressions for probabilities

$$p(00) = \zeta_{00}$$
$$p(10) = \zeta_{00}$$
$$p(01) = \zeta_{00}\zeta_{10}$$
$$p(00) = \zeta_{00}\zeta_{10}\zeta_{11}$$

The corresponding expressions for the moments are

$$1 = \mu_{00} = 2\zeta_{00} + \zeta_{00}\zeta_{10} + \zeta_{00}\zeta_{10}\zeta_{11}$$
$$\mu_{01} = \zeta_{00}\zeta_{10} + \zeta_{00}\zeta_{10}\zeta_{11}$$
$$\mu_{00} = \zeta_{00}\zeta_{10}\zeta_{11}$$

and the normal equations are easily solvable in rational closed form.

We mention without details the following interesting case: sample space of type 2^d and model $x_1, \ldots, x_d, x_1 x_d, x_i x_{i+1}$ for $i = 1, \ldots, d-1$. The case $d = 4$ will be treated in detail in Section 6.9.

6.7 Finitely generated cumulants

From the general theory of exponential models we know that the variances and covariances of the sufficient statistics are in one-to-one correspondence with the model parameters, and could be used to give an alternative set of parameters. In our framework, this relation is of algebraic type. Consider the model

$$p(x_1,\ldots,x_d;\psi) = \exp\left(\sum_{\beta \in L_0'} \psi_\beta x^\beta - K(\psi)\right)$$

with

$$K(\psi) = \ln M(\psi)$$

and

$$M_0(\psi) = \frac{1}{\#D} \sum_{a \in D} \exp\left(\sum_{\beta \in L_0'} \psi_\beta x^\beta\right) \qquad \text{where } D \subseteq (Z_+^d)$$

If we introduce the ζ parameters by

$$\zeta_\beta = e^{\psi_\beta} \qquad \beta \in L_0'$$

we obtain

$$M(\psi) = \frac{1}{\#D} \sum_{a \in D} \left(\prod_{\beta \in L_0'} \zeta_\beta^{a^\beta}\right)$$

and

$$\partial_\beta K(\psi) = \frac{\sum_{a \in D} a^\beta \exp\left(\sum_{\gamma \in L_0'} \psi_\gamma a^\gamma\right)}{\mu(\psi_{L_0'})}$$

There is an algebraic relation between the first derivatives and the second derivatives of $M(\psi_{L_0'})$. In fact,

$$\partial_\beta K(\psi) = \frac{\sum_{a \in D} a^\beta \zeta^{\underline{a}}}{\sum_{a \in D} \zeta^{\underline{a}}}$$

$$\partial_\beta \partial_\gamma K(\psi) = \frac{\sum_{a \in D} a^\beta a^\gamma z^{\underline{a}} - \partial_\beta K(\psi) \partial_\gamma K(\psi)}{\left(\sum_{a \in D} z^{\underline{a}}\right)^2}$$

and

$$\begin{cases} \left(\sum_{a \in D} z^{\underline{a}}\right) \partial_\beta K = \sum_{a \in D} a^\beta z^{\underline{a}} \\ \left(\sum_{a \in D} z^{\underline{a}}\right)^2 \partial_\beta \partial_\gamma K = \sum_{a \in D} a^\beta a^\gamma z^{\underline{a}} - \partial_\beta K \partial_\gamma K \end{cases}$$

Elimination of the relevant indeterminates will show an algebraic relation between the mean parameters and the variance parameters.

Example 78 Consider the uniform distribution on $\{(0,0),(1,0),(0,1)\}$. The exponential model is

$$p(x_1, x_2) = \exp\left(\psi_1 x_1 + \psi_2 x_2 - K(\psi_1, \psi_2)\right)$$

In this case elementary computations show that
$$\begin{cases} \dfrac{\partial^2 K}{\partial \psi_1^2} = \dfrac{\partial K}{\partial \psi_1}\left(1 - \dfrac{\partial K}{\partial \psi_1}\right) \\ \dfrac{\partial^2 K}{\partial \psi_1 \partial \psi_2} = -\dfrac{\partial K}{\partial \psi_1}\dfrac{\partial K}{\partial \psi_2} \\ \dfrac{\partial^2 K}{\partial \psi_2^2} = \dfrac{\partial K}{\partial \psi_2}\left(1 - \dfrac{\partial K}{\partial \psi_2}\right) \end{cases}$$

We want to mention here that a similar property has been considered for statistical models on general sample space under the name of "finite generation of cumulants" in Pistone and Wynn (1999). In the lattice finite sample space, all exponential models have this property.

6.8 Estimating functions

An estimating function is a polynomial $U(\theta, x)$ in the θ parameters and x's indeterminates such that

$$\text{for all } \theta, \quad E_\theta\left(U(\theta, X)\right) = 0 \tag{6.13}$$

The score function is an estimating function; such functions play an important role in the theory of estimation and ancillarity.

If we collect the x's terms in the expression of $U(\theta, x)$

$$U(\theta, X) = \sum_{\alpha \in L} u_\alpha(\theta) X^\alpha$$

then we can write, putting $\mu_\alpha(\theta) = E_\theta(X^\alpha)$,

$$e(\theta) = \sum_{\alpha \in L} u_\alpha(\theta) \mu_\alpha(\theta) = 0$$

As the $\mu_\alpha(\theta)$, $\alpha \in L$ are polynomials of the ring $\mathbb{R}[\theta]$, the computation of the estimating functions can be reduced to the computation of a basis of syzygies, see Cox, Little and O'Shea (1997), formed by S-polynomials, see Definition 24. In the lattice case, we can alternatively look at polynomials in the ζ's parameters.

Equation 6.13 is equivalent to the following condition

$$\begin{aligned} E_\theta\left(U(\theta, X)\right) &= E_0\left(U(\theta, X) p(X, \theta)\right) \\ &= \sum_{\alpha,\beta \in L} u_\alpha \theta_\beta E_0\left(X^\alpha X^\beta\right) \\ &= \sum_{\alpha,\beta \in L} u_\alpha \theta_\beta Q(\alpha, \beta) \\ &= 0 \end{aligned}$$

which provides an alternative algebraic formulation of the problem.

6.9 An extended example

In this example, we discuss a model arising in graphical models, as described in the book by Lauritzen (1996, Chapter 3). This case is simple enough to be treated in few pages, and still it is interesting in showing the application of most of our algebraic methods and introducing the use of toric ideals (Section 6.10).

6.9.1 Sample space

We consider the 2^4 sample space $D = \{0,1\}^4$. It is a case of binary design, with the minimal fan property, already discussed in Section 4.1. We call

$$L = \{\alpha = (\alpha_1, \alpha_2, \alpha_3, \alpha_4) : \alpha_i = 0, 1, \text{ and } i = 1, \ldots, 4\}$$

the list of exponents of the estimable monomials

$$\begin{aligned}\text{Est}(D) &= \{x^\alpha : \alpha \in L\} \\ &= \{1, \\ &\quad x_1, \ x_2, \ x_3, \ x_4, \\ &\quad x_1 x_2, \ x_1 x_3, \ x_1 x_4, \ x_2 x_3, \ x_2 x_4, \ x_3 x_4, \\ &\quad x_1 x_2 x_3, \ x_1 x_2 x_4, \ x_1 x_3 x_4, \\ &\quad x_1 x_2 x_3 x_4\}\end{aligned}$$

which is ordered, with initial ordering

$$x_1 \prec x_2 \prec x_3 \prec x_4$$

by the ordering with matrix

$$\begin{bmatrix} 1 & 1 & 1 & 1 \\ -1 & 0 & 0 & 0 \\ 0 & -1 & 0 & 0 \\ 0 & 0 & -1 & 0 \end{bmatrix}$$

In fact, the first row orders first by degree, while the other rows resolve all ties. Some other monomial ordering give the same ordering of $\text{Est}(D)$.

The design matrix Z is given in Table 6.1, and the matrix of coefficients of indicator polynomials of design points Z^{-1} is given in Table 6.2. Both the matrices Z and Z^{-1} are triangular in the total ordering of monomials. As a second special feature of this binary case, we note that the list of sample points is again L, so that the sample point can be ordered with the same order we used for the list of estimable terms $\text{Est}(D)$.

The matrix of monomial products in the space of random variables $\mathcal{L}(D)$ is given in Table 6.3 to 6.5. In this case, the actual form of the matrix Q is easily described by noting that, on D,

$$X^\alpha X^\beta = X^{\alpha+\beta} = X^{\max(\alpha,\beta)} \tag{6.14}$$

AN EXTENDED EXAMPLE

Table 6.1 Z matrix of the 2^4 sample space D. The last row shows the totals over each monomial in $\mathrm{Est}(D)$.

	1	2	3	4	5	6	7	8	9	10	11	12	13	14	15	16
D	1	x_1	x_2	x_3	x_4	x_1x_2	x_1x_3	x_1x_4	x_2x_3	x_2x_4	x_3x_4	$x_1x_2x_3$	$x_1x_2x_4$	$x_1x_3x_4$	$x_2x_3x_4$	$x_1x_2x_3x_4$
0000	1	0	0	0	0	0	0	0	0	0	0	0	0	0	0	0
1000	1	1	0	0	0	0	0	0	0	0	0	0	0	0	0	0
0100	1	0	1	0	0	0	0	0	0	0	0	0	0	0	0	0
0010	1	0	0	1	0	0	0	0	0	0	0	0	0	0	0	0
0001	1	0	0	0	1	0	0	0	0	0	0	0	0	0	0	0
1100	1	1	1	0	0	1	0	0	0	0	0	0	0	0	0	0
1010	1	1	0	1	0	0	1	0	0	0	0	0	0	0	0	0
1001	1	1	0	0	1	0	0	1	0	0	0	0	0	0	0	0
0110	1	0	1	1	0	0	0	0	1	0	0	0	0	0	0	0
0101	1	0	1	0	1	0	0	0	0	1	0	0	0	0	0	0
0011	1	0	0	1	1	0	0	0	0	0	1	0	0	0	0	0
1110	1	1	1	1	0	1	1	0	1	0	0	1	0	0	0	0
1101	1	1	1	0	1	1	0	1	0	1	0	0	1	0	0	0
1011	1	1	0	1	1	0	1	1	0	0	1	0	0	1	0	0
0111	1	0	1	1	1	0	0	0	1	1	1	0	0	0	1	0
1111	1	1	1	1	1	1	1	1	1	1	1	1	1	1	1	1
Total	16	8	8	8	8	4	4	4	4	4	4	2	2	2	2	1

with $\max(\alpha, \beta) = (\max(\alpha_i, \beta_i); i = 1, 2, 3, 4)$, because all powers are reduced by the equations $x_i^2 = x_i$, $i = 1, 2, 3, 4$.

The inverse matrix Z^{-1} has a special structure, which can be derived in closed form as follows. The polynomial

$$\prod_{i=1}^{4}[(1-a_i) + (2a_i - 1)x_i] = \sum_{\alpha \in L}\left(\prod_{i:a_i=0}(1-a_i)\prod_{i:a_i=1}(2a_i - 1)\right)x^\alpha \quad (6.15)$$

is the indicator polynomial of the sample point

$$a = (a_1, a_2, a_3, a_4)$$

We exploit the fact that the points and the exponents are in the same set L. For each $\alpha < a$ (coordinate-wise), there exists an i such that both $\alpha_i = 0$ and $a_i = 1$, then

$$\alpha < a \implies \prod_{i:\alpha_i=0}(1 - a_i) = 0$$

For $a \leq \alpha$, the value of $\prod_{i:a_i=0}(1 - a_i)$ is 1. The second factor in Equation (6.15)

$$\prod_{i:\alpha_i=1}(2a_i - 1)$$

is a sign depending on the number of i's such that both $\alpha_i = 1$ and $a_i = 0$,

Table 6.2 *Inverse of the Z matrix. The columns are the representations of the indicator functions of the sample points. The columns' labels, e.g, the elements of* Est(D), *are not shown on the table.*

1	2	3	4	5	6	7	8	9	10	11	12	13	14	15	16
0000	1000	0100	0010	0001	1100	1010	1001	0110	0101	0011	1110	1101	1011	0111	1111
1	0	0	0	0	0	0	0	0	0	0	0	0	0	0	0
-1	1	0	0	0	0	0	0	0	0	0	0	0	0	0	0
-1	0	1	0	0	0	0	0	0	0	0	0	0	0	0	0
-1	0	0	1	0	0	0	0	0	0	0	0	0	0	0	0
-1	0	0	0	1	0	0	0	0	0	0	0	0	0	0	0
1	-1	-1	0	0	1	0	0	0	0	0	0	0	0	0	0
1	-1	0	-1	0	0	1	0	0	0	0	0	0	0	0	0
1	-1	0	0	-1	0	0	1	0	0	0	0	0	0	0	0
1	0	-1	-1	0	0	0	0	1	0	0	0	0	0	0	0
1	0	-1	0	-1	0	0	0	0	1	0	0	0	0	0	0
1	0	0	-1	-1	0	0	0	0	0	1	0	0	0	0	0
-1	1	1	1	0	-1	-1	0	-1	0	0	1	0	0	0	0
-1	1	1	0	1	-1	0	-1	0	-1	0	0	1	0	0	0
-1	1	0	1	1	0	-1	-1	0	0	-1	0	0	1	0	0
-1	0	1	1	1	0	0	0	-1	-1	-1	0	0	0	1	0
1	-1	-1	-1	-1	1	1	1	1	1	1	-1	-1	-1	-1	1

Table 6.3 *Matrix* $[Q(\alpha,\beta)]_{\alpha \in L; \beta=1,x_1,x_2,x_3,x_4}$. *To save space, the row labels* Est(D) *are not shown.*

	1	X_1	X_2	X_3	X_4
	1	X_1	X_2	X_3	X_4
	X_1	X_1	X_1X_2	X_1X_3	X_1X_4
	X_2	X_1X_2	X_2	X_2X_3	X_2X_4
	X_3	X_1X_3	X_2X_3	X_3	X_3X_4
	X_4	X_1X_4	X_2X_4	X_3X_4	X_4
	X_1X_2	X_1X_2	X_1X_2	$X_1X_2X_3$	$X_1X_2X_4$
	X_1X_3	X_1X_3	$X_1X_2X_3$	X_1X_3	$X_1X_3X_4$
	X_1X_4	X_1X_4	$X_1X_2X_4$	$X_1X_3X_4$	X_1X_4
	X_2X_4	$X_1X_2X_4$	X_2X_4	$X_2X_3X_4$	X_2X_4
	X_2X_4	$X_1X_2X_4$	X_2X_4	$X_2X_3X_4$	X_2X_4
	X_3X_4	$X_1X_3X_4$	$X_2X_3X_4$	X_3X_4	X_3X_4
	$X_1X_2X_3$	$X_1X_2X_3$	$X_1X_2X_3$	$X_1X_2X_3$	$X_1X_2X_3X_4$
	$X_1X_2X_4$	$X_1X_2X_4$	$X_1X_2X_4$	$X_1X_2X_3X_4$	$X_1X_2X_4$
	$X_1X_3X_4$	$X_1X_3X_4$	$X_1X_2X_3X_4$	$X_1X_3X_4$	$X_1X_3X_4$
	$X_2X_3X_4$	$X_1X_2X_3X_4$	$X_2X_3X_4$	$X_2X_3X_4$	$X_2X_3X_4$
	$X_1X_2X_3X_4$	$X_1X_2X_3X_4$	$X_1X_2X_3X_4$	$X_1X_2X_3X_4$	$X_1X_2X_3X_4$

that is for $\alpha \geq a$ it depends on

$$|\alpha - a| = \sum_i (\alpha_i - a_i)$$

After this reduction, Equation (6.15), becomes

$$\sum_{\alpha \geq a} (-1)^{|\alpha - a|} x^\alpha$$

AN EXTENDED EXAMPLE

Table 6.4 *Matrix* $[Q(\alpha,\beta)]_{\alpha\in L, \beta=x_1x_2, x_1x_3, x_1x_4, x_2x_3, x_2x_4, x_3x_4}$. *To save space, the row labels* Est(D) *are not shown.*

X_1X_2	X_1X_3	X_1X_4	X_2X_3	X_2X_4	X_3X_4
X_1X_2	X_1X_3	X_1X_4	X_2X_3	X_2X_4	X_3X_4
X_1X_2	X_1X_3	X_1X_4	$X_1X_2X_3$	$X_1X_2X_4$	$X_1X_3X_4$
X_1X_2	$X_1X_2X_3$	$X_1X_2X_4$	X_2X_3	X_2X_4	$X_2X_3X_4$
$X_1X_2X_3$	X_1X_3	$X_1X_3X_4$	X_2X_3	$X_2X_3X_4$	X_3X_4
$X_1X_2X_4$	$X_1X_3X_4$	X_1X_4	$X_2X_3X_4$	X_2X_4	X_3X_4
X_1X_2	$X_1X_2X_3$	$X_1X_2X_4$	$X_1X_2X_3$	$X_1X_2X_4$	$X_1X_2X_3X_4$
$X_1X_2X_3$	X_1X_3	$X_1X_3X_4$	$X_1X_2X_3$	$X_1X_2X_3X_4$	$X_1X_3X_4$
$X_1X_2X_4$	$X_1X_3X_4$	X_1X_4	$X_1X_2X_3X_4$	$X_1X_2X_4$	$X_1X_3X_4$
$X_1X_2X_3$	$X_1X_2X_3$	$X_1X_2X_3X_4$	X_2X_3	$X_2X_3X_4$	$X_2X_3X_4$
$X_1X_2X_4$	$X_1X_2X_3X_4$	$X_1X_2X_4$	$X_2X_3X_4$	X_2X_4	$X_2X_3X_4$
$X_1X_2X_3X_4$	$X_1X_3X_4$	$X_1X_3X_4$	$X_2X_3X_4$	$X_2X_3X_4$	X_3X_4
$X_1X_2X_3$	$X_1X_2X_3$	$X_1X_2X_3X_4$	$X_1X_2X_3$	$X_1X_2X_3X_4$	$X_1X_2X_3X_4$
$X_1X_2X_4$	$X_1X_2X_3X_4$	$X_1X_2X_4$	$X_1X_2X_3X_4$	$X_1X_2X_4$	$X_1X_2X_3X_4$
$X_1X_2X_3X_4$	$X_1X_3X_4$	$X_1X_3X_4$	$X_1X_2X_3X_4$	$X_1X_2X_3X_4$	$X_1X_3X_4$
$X_1X_2X_3X_4$	$X_1X_2X_3X_4$	$X_1X_2X_3X_4$	$X_2X_3X_4$	$X_2X_3X_4$	$X_2X_3X_4$
$X_1X_2X_3X_4$	$X_1X_2X_3X_4$	$X_1X_2X_3X_4$	$X_1X_2X_3X_4$	$X_1X_2X_3X_4$	$X_1X_2X_3X_4$

Table 6.5 *Matrix* $[Q(\alpha,\beta)]_{\alpha\in L, \beta=x_1x_2x_3, x_1x_2x_4, x_1x_3x_4, x_2x_3x_4, x_1x_2x_3x_4}$. *To save space, the row labels* Est(D) *are not shown.*

$X_1X_2X_3$	$X_1X_2X_4$	$X_1X_3X_4$	$X_2X_3X_4$	$X_1X_2X_3X_4$
$X_1X_2X_3$	$X_1X_2X_4$	$X_1X_3X_4$	$X_2X_3X_4$	$X_1X_2X_3X_4$
$X_1X_2X_3$	$X_1X_2X_4$	$X_1X_3X_4$	$X_1X_2X_3X_4$	$X_1X_2X_3X_4$
$X_1X_2X_3$	$X_1X_2X_4$	$X_1X_2X_3X_4$	$X_2X_3X_4$	$X_1X_2X_3X_4$
$X_1X_2X_3$	$X_1X_2X_3X_4$	$X_1X_3X_4$	$X_2X_3X_4$	$X_1X_2X_3X_4$
$X_1X_2X_3X_4$	$X_1X_2X_4$	$X_1X_3X_4$	$X_2X_3X_4$	$X_1X_2X_3X_4$
$X_1X_2X_3$	$X_1X_2X_4$	$X_1X_2X_3X_4$	$X_1X_2X_3X_4$	$X_1X_2X_3X_4$
$X_1X_2X_3$	$X_1X_2X_3X_4$	$X_1X_2X_3X_4$	$X_1X_2X_3X_4$	$X_1X_2X_3X_4$
$X_1X_2X_3X_4$	$X_1X_2X_3X_4$	$X_1X_2X_3X_4$	$X_1X_2X_3X_4$	$X_1X_2X_3X_4$
$X_1X_2X_3$	$X_1X_2X_3X_4$	$X_1X_2X_3X_4$	$X_1X_2X_3X_4$	$X_1X_2X_3X_4$
$X_1X_2X_3X_4$	$X_1X_2X_3X_4$	$X_1X_2X_3X_4$	$X_1X_2X_3X_4$	$X_1X_2X_3X_4$
$X_1X_2X_3X_4$	$X_1X_2X_3X_4$	$X_1X_2X_3X_4$	$X_1X_2X_3X_4$	$X_1X_2X_3X_4$
$X_1X_2X_3$	$X_1X_2X_3X_4$	$X_1X_2X_3X_4$	$X_1X_2X_3X_4$	$X_1X_2X_3X_4$
$X_1X_2X_3X_4$	$X_1X_2X_4$	$X_1X_2X_3X_4$	$X_1X_2X_3X_4$	$X_1X_2X_3X_4$
$X_1X_2X_3X_4$	$X_1X_2X_3X_4$	$X_1X_2X_3X_4$	$X_1X_2X_3X_4$	$X_1X_2X_3X_4$
$X_1X_2X_3X_4$	$X_1X_2X_3X_4$	$X_1X_2X_3X_4$	$X_2X_3X_4$	$X_1X_2X_3X_4$
$X_1X_2X_3X_4$	$X_1X_2X_3X_4$	$X_1X_2X_3X_4$	$X_1X_2X_3X_4$	$X_1X_2X_3X_4$

as it can be checked in Table 6.2. Given any real function f on the sample space D, the interpolation formula is

$$f(x) = \sum_{a\in D} f(a) \sum_{\alpha\geq a} (-1)^{|\alpha-a|} x^\alpha$$
$$= \sum_{\alpha\in L} \left(\sum_{a\leq\alpha} (-1)^{|\alpha-a|} f(a) \right) x^\alpha \qquad (6.16)$$

The reader may recognise this a special case of Möbius inversion.

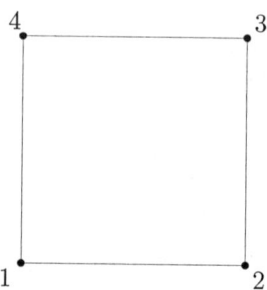

Figure 6.1 *A graphical model.*

6.9.2 The model

We consider an exponential model of the form

$$p(x;\phi) = \exp\left(\sum_{\alpha \in M} \psi_\alpha x^\alpha\right)$$

where

$$M = \{0000, 1000, 0100, 0010, 0001, 1100, 1001, 0110, 0011\} \subset L$$

and $\psi = (\psi_\alpha)_{\alpha \in M_0}$, $M_0 = M \setminus \{0000\}$. Note that this model is a submodel of the full exponential model and that its monomial terms are not the first in the ordering we are using. In fact, the rank of the terms in M is 1, 2, 3, 4, 5, 6, 8, 9, 11. This model is described as a graphical model in Figure 6.1 and is characterised by its Markov properties. For example, the factorisation

$$\begin{aligned}p(x;\psi) &= \exp(\psi_{0000} + \psi_{0100}x_2 + \psi_{0001}x_4)\\ &\quad \exp(\psi_{1000}x_1 + \psi_{1100}x_1x_2 + \psi_{1001}x_1x_4)\\ &\quad \exp(\psi_{0010}x_3 + \psi_{0110}x_2x_3 + \psi_{0011}x_3x_4)\end{aligned}$$

shows the conditional independence of X_1 and X_3, given X_2, X_4, see Lauritzen (1996, Chapter 3). Similarly X_2 and X_4 are conditionally independent given X_1, X_3.

Let us consider the same model as a density P_θ in polynomial representation:

$$P_\theta = \theta_{0000} + \sum_{\alpha \in L_0} \theta_\alpha X^\alpha \qquad (6.17)$$

with the normalising condition

$$\theta_{0000} = 1 - \sum_{\alpha \in L_0} \theta_\alpha m_\alpha$$

where $E_0(X^\alpha) = m_\alpha = t_\alpha/16$ is the α-moment with respect to the uniform

AN EXTENDED EXAMPLE

Table 6.6 *Linear transformation from the θ parameters to the μ parameters* ($\times 16$).

M	L	0000	1000	0100	0010	0001	1100	1010	1001	0110	0101	0011	1110	1101	1011	0111	1111
1	0000	16	8	8	8	8	4	4	4	4	4	4	2	2	2	2	1
2	1000	8	8	4	4	4	4	4	4	2	2	2	2	2	2	1	1
3	0100	8	4	8	4	4	4	2	2	4	4	2	2	2	1	2	1
4	0010	8	4	4	8	4	2	4	2	4	2	4	2	1	2	2	1
5	0001	8	4	4	4	8	2	2	4	2	4	4	1	2	2	2	1
6	1100	4	4	4	2	2	4	2	2	2	2	1	2	2	1	1	1
	1010	4	4	2	4	2	2	4	2	2	1	2	2	1	2	1	1
8	1001	4	4	2	2	4	2	2	4	1	2	2	1	2	2	1	1
9	0110	4	2	4	4	2	2	2	1	4	2	2	2	1	1	2	1
	0101	4	2	4	2	4	2	1	2	2	4	2	1	2	1	2	1
11	0011	4	2	2	4	4	1	2	2	2	2	4	1	1	2	2	1
	1110	2	2	2	2	1	2	2	1	2	1	1	2	1	1	1	1
	1101	2	2	2	1	2	2	1	2	1	2	1	1	2	1	1	1
	1011	2	2	1	2	2	1	2	2	1	1	2	1	1	2	1	1
	0111	2	1	2	2	2	1	1	1	2	2	2	1	1	1	2	1
	1111	1	1	1	1	1	1	1	1	1	1	1	1	1	1	1	1

distribution, and $(t_\alpha; \alpha \in L)$ is a vector of totals over $\mathrm{Est}(D)$, in the last row of Table 6.2.

The expectation parameters

$$\mu_\beta = \sum_{a \in D} X^\beta(a) P_\theta(a)$$

can be computed as a function of the θ parameters by using the Q-matrix or directly from Equation (6.14) as follows

$$\begin{aligned}\mu_\beta &= E_\theta(X^\beta) \\ &= E_0(X^\beta P_\theta) \\ &= \sum_{\alpha \in L} \theta_\alpha E_0(X^\beta X^\alpha) \\ &= \sum_{\alpha \in L} \theta_\alpha m_{\max(\beta,\alpha)}\end{aligned}$$

We know that the expectation parameters μ_β, $\beta \in M_0$ are a full parametrisation of our exponential model. The matrix with entries $16 \times [m_{\max(\beta,\alpha)}]$, for $\beta \in M$ and $\alpha \in L$ is given in Table 6.6. The relevant part is a matrix $(M \times L)$ rectangular because the μ parameters are restricted only by the normalisation condition $\mu_{0000} = 1$, while the θ parameters belong to an eight-dimensional sub-variety of $\mathbb{R}^{16}$, to be described below.

6.9.3 Lattice case

As the sample space is integer valued, it is possible to introduce one more parameterisation, namely

$$p(x;\psi) = \exp\left(\sum_{\alpha \in M} \psi_\alpha x^\alpha\right)$$
$$= \prod_{\alpha \in M} \exp(\psi_\alpha x^\alpha) \qquad (6.18)$$
$$= \zeta_0 \prod_{\alpha \in M_0} \zeta_\alpha^{x^\alpha}$$

where

$$\zeta_\alpha = \exp(\psi_\alpha), \quad \alpha \in M_0$$

As in Equation (6.17) the polynomial form of the density as a function of the ζ parameters can be computed in at least two ways. One possibility is to write the polynomial interpolation of each factor in Equation (6.18) and then collect the terms in

$$\exp(\psi_\alpha x^\alpha) = \zeta_\alpha^{x^\alpha} = 1 + (\zeta_\alpha - 1)x^\alpha$$

and

$$\zeta_0 \prod_{\alpha \in M_0} \zeta_\alpha^{x^\alpha} = \zeta_0 \prod_{\alpha \in M_0} (1 + (\zeta_\alpha - 1)x^\alpha)$$

Alternatively, one could use directly the generic interpolation formula of Equation (6.16) and obtain

$$\theta_\alpha = \zeta_0 \sum_{a \leq \alpha} (-1)^{|\alpha - a|} \prod_{\alpha \in M_0} \zeta_\alpha^{a^\alpha}$$

for $\alpha \in L$. The coefficients in the previous equations are the same as the entries of the matrix $(Z^{-1})^t$, but the new system is not linear: the monomial in each term has its exponent equal to a row of the matrix Z. The monomials associated with each row of Table 6.1 are listed below.

$$\begin{aligned}
&0000 \mapsto p_1 = \zeta_0 &&\quad 0110 \mapsto p_9 = \zeta_0\zeta_2\zeta_3\zeta_7 \\
&1000 \mapsto p_2 = \zeta_0\zeta_1 &&\quad 0101 \mapsto p_{10} = \zeta_0\zeta_2\zeta_4 \\
&0100 \mapsto p_3 = \zeta_0\zeta_2 &&\quad 0011 \mapsto p_{11} = \zeta_0\zeta_3\zeta_4\zeta_8 \\
&0010 \mapsto p_4 = \zeta_0\zeta_3 &&\quad 1110 \mapsto p_{12} = \zeta_0\zeta_1\zeta_2\zeta_3\zeta_5\zeta_7 \\
&0001 \mapsto p_5 = \zeta_0\zeta_4 &&\quad 1101 \mapsto p_{13} = \zeta_0\zeta_1\zeta_2\zeta_4\zeta_5\zeta_6 \\
&1100 \mapsto p_6 = \zeta_0\zeta_1\zeta_2\zeta_5 &&\quad 1011 \mapsto p_{14} = \zeta_0\zeta_1\zeta_3\zeta_4\zeta_6\zeta_8 \\
&1010 \mapsto p_7 = \zeta_0\zeta_1\zeta_3 &&\quad 0111 \mapsto p_{15} = \zeta_0\zeta_2\zeta_3\zeta_4\zeta_7\zeta_8 \\
&1001 \mapsto p_8 = \zeta_0\zeta_1\zeta_4\zeta_6 &&\quad 1111 \mapsto p_{16} = \zeta_0\zeta_1\zeta_2\zeta_3\zeta_4\zeta_5\zeta_6\zeta_7\zeta_8
\end{aligned} \qquad (6.19)$$

We write explicitly all the equations relating the θ parameters to the ζ

AN EXTENDED EXAMPLE

parameters as follows:

$$\begin{cases} \theta_{0000} = \zeta_0 \\ \theta_{1000} = \zeta_0(\zeta_1 - 1) \\ \theta_{0100} = \zeta_0(\zeta_2 - 1) \\ \theta_{0010} = \zeta_0(\zeta_3 - 1) \\ \theta_{0001} = \zeta_0(\zeta_4 - 1) \\ \theta_{1100} = \zeta_0(\zeta_1\zeta_2\zeta_5 - \zeta_1 - \zeta_2 + 1) \\ \theta_{1010} = \zeta_0(\zeta_1\zeta_3 - \zeta_1 - \zeta_3 + 1) \\ \theta_{1001} = \zeta_0(\zeta_1\zeta_4\zeta_6 - \zeta_1 - \zeta_4 + 1) \\ \theta_{0110} = \zeta_0(\zeta_2\zeta_3\zeta_7 - \zeta_2 - \zeta_3 + 1) \\ \theta_{0101} = \zeta_0(\zeta_2\zeta_4 - \zeta_2 - \zeta_4 + 1) \\ \theta_{0011} = \zeta_0(\zeta_3\zeta_4\zeta_8 - \zeta_3 - \zeta_4 + 1) \\ \theta_{1110} = \zeta_0(\zeta_1\zeta_2\zeta_3\zeta_5\zeta_7 - \zeta_1\zeta_2\zeta_5 - \zeta_2\zeta_3\zeta_7 - \zeta_1\zeta_3 + \zeta_1 + \zeta_2 + \zeta_3 - 1) \\ \theta_{1101} = \zeta_0(\zeta_1\zeta_2\zeta_4\zeta_5\zeta_6 - \zeta_1\zeta_2\zeta_5 - \zeta_1\zeta_4\zeta_6 - \zeta_2\zeta_4 + \zeta_1 + \zeta_2 + \zeta_4 - 1) \\ \theta_{1011} = \zeta_0(\zeta_1\zeta_3\zeta_4\zeta_6\zeta_8 - \zeta_1\zeta_4\zeta_6 - \zeta_3\zeta_4\zeta_8 - \zeta_1\zeta_3 + \zeta_1 + \zeta_3 + \zeta_4 - 1) \\ \theta_{0111} = \zeta_0(\zeta_2\zeta_3\zeta_4\zeta_7\zeta_8 - \zeta_2\zeta_3\zeta_7 - \zeta_3\zeta_4\zeta_8 - \zeta_2\zeta_4 + \zeta_2 + \zeta_3 + \zeta_4 - 1) \\ \theta_{1111} = \zeta_0(\zeta_1\zeta_2\zeta_3\zeta_4\zeta_5\zeta_6\zeta_7\zeta_8 - \zeta_1\zeta_2\zeta_4\zeta_5\zeta_6 - \zeta_1\zeta_2\zeta_3\zeta_5\zeta_7 \\ \qquad\quad - \zeta_1\zeta_3\zeta_4\zeta_6\zeta_8 - \zeta_2\zeta_3\zeta_4\zeta_7\zeta_8 + \zeta_1\zeta_2\zeta_5 \\ \qquad\quad + \zeta_1\zeta_4\zeta_6 + \zeta_2\zeta_3\zeta_7 + \zeta_3\zeta_4\zeta_8 + \zeta_1\zeta_3 \\ \qquad\quad + \zeta_2\zeta_4 - \zeta_1 - \zeta_2 - \zeta_3 - \zeta_4 + 1) \end{cases}$$
(6.20)

An equivalent system can be obtained by equating at each sample point the two presentations of the probability density, that is, by writing the equations

$$p(a, \theta) = \zeta_0 \prod_{\alpha \in M_0} \zeta_\alpha^{a^\alpha}, \quad a \in D$$

with $p = Z\theta$. In our example we obtain Equation (6.21).

The elimination of the ζ variables in either Equation (6.20) or Equation (6.21), gives the equations describing the variety of the values of the θ parameters compatible with the model. This elimination is relatively computationally heavy, so it is simpler to proceed using the parameters given by the values of the probability at each point, that is, using Equation (6.19). As will be discussed in Section 6.10, in this case we have a homogeneous binomial ideal, namely each polynomial has only two terms of the same degree.

$$\begin{cases} \zeta_0 = \theta_1 \\ \zeta_0\zeta_1 = \theta_1 + \theta_2 \\ \zeta_0\zeta_2 = \theta_1 + \theta_3 \\ \zeta_0\zeta_3 = \theta_1 + \theta_4 \\ \zeta_0\zeta_4 = \theta_1 + \theta_5 \\ \zeta_0\zeta_1\zeta_3 = \theta_1 + \theta_2 + \theta_3 + \theta_6 \\ \zeta_0\zeta_1\zeta_3 = \theta_1 + \theta_2 + \theta_4 + \theta_7 \\ \zeta_0\zeta_1\zeta_4\zeta_6 = \theta_1 + \theta_2 + \theta_5 + \theta_8 \\ \zeta_0\zeta_2\zeta_3\zeta_7 = \theta_1 + \theta_3 + \theta_4 + \theta_9 \\ \zeta_0\zeta_2\zeta_4 = \theta_1 + \theta_3 + \theta_5 + \theta_{10} \\ \zeta_0\zeta_3\zeta_4\zeta_8 = \theta_1 + \theta_4 + \theta_5 + \theta_{11} \\ \zeta_0\zeta_1\zeta_2\zeta_3\zeta_5\zeta_7 = \theta_1 + \theta_2 + \theta_3 + \theta_4 + \theta_6 + \theta_7 + \theta_9 + \theta_{12} \\ \zeta_0\zeta_1\zeta_2\zeta_4\zeta_5\zeta_6 = \theta_1 + \theta_2 + \theta_3 + \theta_5 + \theta_6 + \theta_8 + \theta_{10} + \theta_{13} \\ \zeta_0\zeta_1\zeta_3\zeta_4\zeta_6\zeta_8 = \theta_1 + \theta_2 + \theta_4 + \theta_5 + \theta_7 + \theta_8 + \theta_{11} + \theta_{14} \\ \zeta_0\zeta_2\zeta_3\zeta_4\zeta_7\zeta_8 = \theta_1 + \theta_3 + \theta_4 + \theta_5 + \theta_9 + \theta_{10} + \theta_{11} + \theta_{15} \\ \zeta_0\zeta_1\zeta_2\zeta_3\zeta_4\zeta_5\zeta_6\zeta_7\zeta_8 = \theta_1 + \theta_2 + \theta_3 + \theta_4 + \theta_5 + \theta_6 + \theta_7 + \theta_8 + \theta_9 \\ \qquad\qquad\qquad\qquad\qquad + \theta_{10} + \theta_{11} + \theta_{12} + \theta_{13} + \theta_{14} + \theta_{15} + \theta_{16} \end{cases} \quad (6.21)$$

Direct elimination of the ζ's leads to

$$\begin{cases} -p_3p_5 + p_1p_{10} = 0 & p_1p_8p_{10}p_{12} + p_2p_5p_9p_{13} = 0 \\ -p_2p_4 + p_1p_7 = 0 & -p_6p_8 + p_2p_{13} = 0 \\ -p_9p_{11} + p_4p_{15} = 0 & p_3p_7p_{11}p_{13} - p_4p_6p_{10}p_{14} = 0 \\ p_5p_9p_{14} - p_4p_8p_{15} = 0 & p_4p_8p_{10}p_{12} - p_5p_7p_9p_{13} = 0 \\ -p_1p_9p_{10}p_{14} + p_3p_4p_8p_{15} = 0 & -p_5p_6p_{14}p_{15} + p_2p_{10}p_{11}p_{16} = 0 \\ -p_8p_{11} + p_5p_{14} = 0 & -p_4p_6p_{14}p_{15} + p_3p_7p_{11}p_{16} = 0 \\ p_2p_9p_{10}p_{14} - p_3p_7p_8p_{15} = 0 & -p_1p_6p_{14}p_{15} + p_2p_3p_{11}p_{16} = 0 \\ p_3p_{11}p_{12} - p_4p_6p_{15} = 0 & -p_{13}p_{15} + p_{10}p_{16} = 0 \\ -p_1p_{10}p_{11}p_{12} + p_4p_5p_6p_{15} = 0 & p_4p_8p_{12}p_{15} - p_5p_7p_9p_{16} = 0 \\ -p_6p_9 + p_3p_{12} = 0 & p_2p_9p_{13}p_{14} - p_3p_7p_8p_{16} = 0 \\ p_2p_{10}p_{11}p_{12} - p_5p_6p_7p_{15} = 0 & p_1p_8p_{12}p_{15} - p_2p_5p_9p_{16} = 0 \\ p_2p_{11}p_{13} - p_5p_6p_{14} = 0 & p_1p_9p_{13}p_{14} - p_3p_4p_8p_{16} = 0 \\ -p_1p_7p_{11}p_{13} + p_4p_5p_6p_{14} = 0 & p_1p_{11}p_{12}p_{13} - p_4p_5p_6p_{16} = 0 \\ -p_3p_8p_{12} + p_2p_9p_{13} = 0 & -p_{12}p_{14} + p_7p_{16} = 0 \end{cases} \quad (6.22)$$

The corresponding variety for the θ and η parameters can be obtained by substitution into this system of equations. Note that the representation in Equation (6.22) is not unique. We return to this point in Section 6.10.

6.9.4 Maximum likelihood

The normal equations for estimating parameter values can be based in an exponential model on sample values $\bar{\mu}_\beta$, $\beta \in M$, namely

$$\bar{\mu}_\beta = \mathrm{E}_\zeta \left(X^\beta \right), \qquad \beta \in M_0$$

By use of the expression of μ parameters as a function of ζ's parameters given by Equation (6.20), we get, by multiplying the matrix $Z_1 = Z_{\alpha \in L, \beta \in M}$ by the vector of monomials representing the p-values as a function of the ζ-parameters, the normal equations

$$Z_1^t [p] = \bar{\mu}_\beta$$

Since the ζ-parameters are an algebraically free set, this can be considered the most compact expression.

6.10 Orthogonality and toric ideals

The central role played by the Z matrix for the full model, in any of the examples, and familiarity with experimental design encourages the use of orthogonality to aid the different parametric representation developed in the example of Section 6.9.

Without loss of generality we can replace Z by the partitioned matrix

$$\tilde{Z} = [Z_1 : Z_2]$$

where Z_1 (as above) is the restriction of Z to the model columns, and Z_2 is full rank and has column space orthogonal to the column space of Z_1. For the example in Section 6.9 this is straightforward: simply replace the $0, 1$ coding of levels by the $-1, +1$ coding, and generate Z_2 accordingly. Listing of the non-model terms in the order $x_1 x_3$, $x_2 x_4$, $x_1 x_2 x_3$, $x_1 x_2 x_4$, $x_1 x_3 x_4$, $x_2 x_3 x_4$, $x_1 x_2 x_3 x_4$ one obtains the 16×7 matrix for Z_2 in Table 6.7. Then, using the original coding for Z, it remains to see that Z_1 and Z_2 are orthogonal: $Z_1^t Z_2 = 0$. This depends on the fact that the model part in the new coding is orthogonal to Z_1, and the recoding does not change the column space. Using the notation $[p]$, $[\psi]$, and $[\hat{\mu}]$ to denote the vector quantities as column vectors in the model, we have a succinct representation for the model

$$[p] = \exp \left(Z_1 [\psi] \right)$$

or

$$\log [p] = Z_1 [\psi]$$

The last equation is equivalent to

$$Z_2^t \log [p] = 0 \qquad (6.23)$$

from the orthogonality of Z_1 and Z_2.

Table 6.7 *Matrix Z_2 for the graphical model of Section 6.9.*

	x_1x_3	x_2x_4	$x_1x_2x_3$	$x_1x_2x_4$	$x_1x_3x_4$	$x_2x_3x_4$	$x_1x_2x_3x_4$
0000	1	1	−1	−1	−1	−1	1
1000	−1	1	1	1	1	−1	−1
0100	1	−1	1	1	−1	1	−1
0010	−1	1	1	−1	1	1	−1
0001	1	−1	−1	1	1	1	−1
1100	−1	−1	−1	−1	1	1	1
1010	1	1	−1	1	−1	1	1
1001	−1	−1	1	−1	−1	1	1
0110	−1	−1	−1	1	1	−1	1
0101	1	1	1	−1	1	−1	1
0011	−1	−1	1	1	−1	−1	1
1110	1	−1	1	−1	−1	−1	−1
1101	−1	1	−1	1	−1	−1	−1
1011	1	−1	−1	−1	1	−1	−1
0111	−1	1	−1	−1	−1	1	−1
1111	1	1	1	1	1	1	1

Let us consider Equation (6.23) in some detail. Suppose that all the entries of p are positive. In most cases Z_2 will have integer entries; strictly if Z_1 has integer entries, as in the lattice case, then an orthogonal Z_2 with integer entries can be computed. In the example the entries are 0, 1. Moreover, since Est always contains the constant term and if the model is chosen to include the constant term (the ψ_0 term), then we have that $Z_2^t 1 = 0$, where $1 = [1, 1, \ldots, 1]^t$. These two facts together show that Equation (6.23) lead to an homogeneous toric ideal for the p-parameters.

Consider the example, then for $[p] > 0$, the equation $Z_2^t \log[p] = 0$ implies a bank of equations for the p-parameters

$$\begin{cases} p_1 p_3 p_5 p_7 p_{10} p_{12} p_{14} p_{16} - p_2 p_4 p_6 p_8 p_9 p_{11} p_{13} p_{15} = 0 \\ p_1 p_2 p_4 p_7 p_{10} p_{13} p_{15} p_{16} - p_3 p_5 p_6 p_8 p_9 p_{11} p_{12} p_{14} = 0 \\ p_2 p_3 p_4 p_8 p_{10} p_{11} p_{12} p_{16} - p_1 p_5 p_6 p_7 p_9 p_{13} p_{14} p_{15} = 0 \\ p_2 p_3 p_5 p_7 p_9 p_{11} p_{13} p_{16} - p_1 p_4 p_6 p_8 p_{10} p_{12} p_{14} p_{15} = 0 \\ p_2 p_4 p_5 p_6 p_9 p_{10} p_{14} p_{16} - p_1 p_3 p_7 p_8 p_{11} p_{12} p_{13} p_{15} = 0 \\ p_3 p_4 p_5 p_6 p_7 p_8 p_{15} p_{16} - p_1 p_2 p_9 p_{10} p_{11} p_{12} p_{13} p_{14} = 0 \\ p_1 p_6 p_7 p_8 p_9 p_{10} p_{11} p_{16} - p_2 p_3 p_4 p_5 p_{12} p_{13} p_{14} p_{15} = 0 \end{cases} \quad (6.24)$$

This provides an alternative set of conditions to those obtained by elimination in Equation (6.22). Simpler versions are obtained by a column reduction of Z_2. The two representations agree when $[p] > 0$, namely all probabilities are positive on the support.

With regard to the maximum likelihood estimators, it can be shown that: if there is such an estimator $[\hat{p}] > 0$, then this is the only one satisfying positivity. Thus,

1. if there is a solution to the normal equations $[\hat{p}] > 0$, then it lies on the intersection of the linear variety $Z_1^t[p] = \overline{[\mu]}$ and the variety derived from $Z_2^t \log[p] = 0$.

2. Such a solution can be found as one solution of the algebraic normal equations expressed in any of the representations using one of p, ζ, θ, μ.

The theory of toric ideals lies behind the representations (6.19), (6.22), (6.23) and (6.24) in the example. A full development is beyond the scope of the book, but the reader who has reached this point will have found several of the main features of the theory. In the book by Kreuzer and Robbiano (2000), there is a pleasing introduction to the theory of toric ideals. They were first used in statistics in the work of Diaconis and Sturmfels (1998), see also Dinwoodie (1998). This final section can be considered as a brief introduction to this important development, showing the critical role of the ζ parametrisation.

The representation of probabilities in terms of the ζ in (6.19) is referred to as a representation in terms of "power products". For distributions and models supported on more complex integer lattice points, this power product representation still holds. Now, power products always lie on a toric ideal generated by special homogeneous binary expressions. *Binary* here means having only two terms. It is important to note that this binary representation is not unique, and one may ask for a minimal set of such generators, that is, a minimal basis. A simple way to obtain a basis is precisely by elimination, which is how (6.19) was found.

The linear representation (6.23) for $\log[p]$ which depends on the matrix Z_2, orthogonal to Z_1, can also be found in the theory. Writing $[q] = \log[p]$ the equation is $Z_2^t[q] = 0$. All scalar multiples of Z_1 also satisfy this equation and such integer solutions form an integer lattice. Column-reduced versions generate the same lattice. This is a way of constructing bases for the toric ideal with fewer generators than simple elimination.

What is perhaps particular about the development in this book is the demonstration of the construction of the toric ideal for any discrete probability model on lattice points, L, arising out of the polynomial exponential representation. We summarize this as a series of steps.

1. From a support D on integer lattice points and a monomial ordering τ, construct $\text{Est}_\tau = \{x^\alpha : \alpha \in L\}$.

2. Interpolate the logarithm of the probability function on D to give a saturated exponential model given by L.

3. Take any submodel $L' \subset L$ and parameterise the exponential terms with $\zeta = \exp(\psi)$.

4. Claim that the probabilities p are power products in the ζ.

5. Construct the toric ideal in the p's by elimination.

6. Construct a matrix Z_2, orthogonal to Z_1, and use it, or use the toric ideal theory, to construct a minimal set of generators for the ideal in Item 5.

We can restate the formulation of the maximum likelihood problem. If the maximum likelihood estimator $[\hat{p}]$ satisfies $[\hat{p}] > 0$ then it lies on the intersection of the linear (affine) variety given by the normal equations and the toric variety defined by the submodel.

The following provides an easy exercise.

Example 79 Consider a 3×3 contingency table with levels coded $-1, 0, +1$ but with the diagonal $x_1 = x_2$ missing. Thus

$$D = \{(0, -1), (1, -1), (1, 0), (0, 1), (-1, 1), (-1, 0)\}$$

and label the corresponding probabilities $p_1, \ldots, p_6$, respectively. For any graded ordering, τ, such as tdeg

$$\text{Est}_\tau = \{1,\ x_1,\ x_2,\ x_1^2,\ x_2^2,\ x_1 x_2\}$$

Consider the exponential submodel excluding the $x_1 x_2$ term:

$$p(x; \psi) = \exp\left(\psi_0 + \psi_1 x_1 + \psi_2 x_2 + \psi_3 x_1^2 + \psi_4 x_2^2\right)$$

Then, in the above development

$$Z_1 = \begin{bmatrix} 1 & 0 & -1 & 0 & 1 \\ 1 & 1 & -1 & 1 & 1 \\ 1 & 1 & 0 & 1 & 0 \\ 1 & 0 & 1 & 0 & 1 \\ 1 & -1 & 1 & 1 & 1 \\ 1 & -1 & 0 & 1 & 0 \end{bmatrix},\ Z_2 = \begin{bmatrix} 1 \\ -1 \\ 1 \\ -1 \\ 1 \\ -1 \end{bmatrix}$$

The toric ideal is generated by

$$p_1 p_3 p_5 = p_2 p_4 p_6$$

ORTHOGONALITY AND TORIC IDEALS

and the full set of maximum likelihood equations is

$$\hat{p}_1 + \hat{p}_2 + \hat{p}_3 + \hat{p}_4 + \hat{p}_5 + \hat{p}_6 = 1$$
$$\hat{p}_2 + \hat{p}_3 - \hat{p}_5 - \hat{p}_6 = \bar{\mu}_{10}$$
$$-\hat{p}_1 - \hat{p}_2 + \hat{p}_4 + \hat{p}_5 = \bar{\mu}_{01}$$
$$\hat{p}_2 + \hat{p}_3 + \hat{p}_5 + \hat{p}_6 = \bar{\mu}_{20}$$
$$\hat{p}_1 + \hat{p}_2 + \hat{p}_4 + \hat{p}_5 = \bar{\mu}_{02}$$
$$\hat{p}_1 \hat{p}_3 \hat{p}_5 - \hat{p}_2 \hat{p}_4 \hat{p}_6 = 0$$

The solution has "closed form" in which, for example $\hat{p}_1$ is a solution of a cubic equation and $\hat{p}_2, \ldots, \hat{p}_6$ are elementary functions of this solution.

References

Abbott J., Bigatti A., Kreutzer M., and Robbiano L. (2000) Computing ideals of points. *J. Symb. Comput.* **30**, 341–356.

Adams W.W. and Loustaunau P. (1994) *An Introduction to Gröbner Bases*. American Mathematical Society, Providence, RI.

Amari S.i. (1985) *Differential-geometrical Methods in Statistics*. Springer-Verlag, New York, 2nd printing 1990 corrected ed.

Andrews D.F. and Stafford J.E. (2000) *Symbolic Computation for Statistical Inference*. Oxford University Pres, Oxford.

Anthony M. and Biggs N. (1997) *Computational Learning Theory. An Introduction*. Cambridge University Press, Cambridge.

Barlow R.E. (1998) *Engineering Reliability*. Society for Industrial and Applied Mathematics (SIAM), Philadelphia, PA.

Barndorff-Nielsen O.E. and Cox D.R. (1989) *Asymptotic Techniques for Use in Statistics*. Chapman & Hall, London.

Barndorff-Nielsen O.E. and Cox D.R. (1994) *Inference and Asymptotics*. Chapman & Hall, London.

Becker T. and Weispfenning V. (1993) *Gröbner Bases. A Computational Approach to Commutative Algebra*. Springer-Verlag, New York.

Box G.E.P., Hunter W.G., and Hunter J.S. (1978) *Statistics for Experimenters*. John Wiley & Sons, New York-Chichester-Brisbane.

Buchberger B. (1966) *On Finding a Vector Space Basis of the Residue Class Ring Modulo a Zero Dimensional Polynomial Ideal*. PhD thesis, Department of Mathematics, University of Innsbruck.

Caboara M., de Dominicis G., and Robbiano L. (1996) Multigraded Hilbert function and Buchberger algorithm. In Y. Lakshman, (ed.) *Proc. ISSAC'96*, ACM.

Caboara M., Pistone G., Riccomagno E., and Wynn H. (1997) The fan of an experimental design. SCU Research Report 33, Department of Statistics, University of Warwick.

Caboara M. and Riccomagno E. (1998) An algebraic computational approach to the identifiability of Fourier models. *J. Symbolic Comput.* **26**, 245–260.

Caboara M. and Robbiano L. (1997) Families of ideals in Statistics. In Küchlin, (ed.) *Proc. ISSAC '97*, ACM Press, New York.

Collart S., Kalkbrener M., and Mall D. (1997) Converting bases with the Gröbner walk. *J. Symbolic Comput.* **24**, 465–469.

Cox D., Little J., and O'Shea D. (1997) *Ideals, Varieties, and Algorithms*. Springer-Verlag, New York, 2nd ed.

Cox D., Little J., and O'Shea D. (1998) *Using Algebraic Geometry*. Springer-Verlag, New York.

REFERENCES

Cox D.R. and Reid N. (2000) *The Theory of the Design of Experiments*. Chapman & Hall, London.

Diaconis P. and Sturmfels B. (1998) Algebraic algorithms for sampling from conditional distributions. *Ann. Statist.* **26**, 363–397.

Dinwoodie I.H. (1998) The Diaconis-Sturmfels algorithm and rules of succession. *Bernoulli* **4**, 401–410.

Dohmen K. (1999) Improved inclusion-exclusion identities and inequalities based on a particular class of abstract tubes. *Electron. J. Probab.* **4**, no. 5, 12 pp. (electronic).

Faugère J.C., Gianni P., Lazard D., and Mora T. (1993) Efficient computation of zero-dimensional Gröbner bases by change of ordering. *J. Symbolic Comput.* **16**, 329–344.

Fontana R., Pistone G., and Rogantin M.P. (1997) Algebraic analysis and generation of two-levels designs. *Statistica Applicata* **9**, 15–29.

Fontana R., Pistone G., and Rogantin M.P. (2000) Classification of two-level factorial fractions. *J. Statist. Plann. Inference* **87**, 149–172.

Giglio B., Naiman D.Q., and Wynn H.P. (2000) Gröbner bases, abstract tubes, and inclusion-exclusion reliability bounds. SCU Research Report 25, Department of Statistics, Uuniversity of Warwick.

Giglio B., Riccomagno E., and Wynn H.P. (2000) Gröbner bases strategies in regression. *J. Appl. Statist.* **27**, 923–938.

Halmos P. and Givant S. (1998) *Logic as Algebra*. Mathematical Association of America, Washington, DC.

Holliday T., Pistone G., Riccomagno E., and Wynn H.P. (1999) The application of computational algebraic geometry to the analysis of designed experiments: a case study. *Comput. Statist.* **14**, 213–231.

Kiefer J.C. (1987) *Introduction to Statistical Inference. Edited and with a preface by Gary Lorden*. Springer-Verlag, New York.

Kreuzer M. and Robbiano L. (2000) *Computational Commutative Algebra 1*. Springer, Berlin-Heidelberg.

Lauritzen S.L. (1996) *Graphical Models*. The Clarendon Press Oxford University Press, New York.

Lehmann E.L. (1983) *Theory of Point Estimation*. John Wiley & Sons Inc., New York, a Wiley Publication in Mathematical Statistics.

Lehmann E.L. (1986) *Testing Statistical Hypotheses*. John Wiley & Sons Inc., New York, second ed.

Marinari M., Möller H., and Mora T. (1996) Gröbner duality. Tech. Rep. 312, Università di Genova, Dipartimento di Matematica, preprint.

Marinari M.G., Möller H.M., and Mora T. (1993) Gröbner bases of ideals defined by functionals with an application to ideals of projective points. *Appl. Algebra Engrg. Comm. Comput.* **4**, 103–145.

McCullagh P. and Nelder J.A. (1983) *Generalized Linear Models*. Chapman & Hall, London.

Mora T. (1994) An introduction to commutative and noncommutative Gröbner bases. *Theoret. Comput. Sci.* **134**, 131–173.

Mora T. and Robbiano L. (1988) The Gröbner fan of an ideal. *J. Symbolic Comput.* **6**, 183–208.

REFERENCES

Naiman D. and Wynn H.P. (2000) Abstract tubes for simplex and orthant arrangements with applications to reliability bounds. SCU Research Report 24, Department of Statistics, Uuniversity of Warwick.

Pistone G. and Wynn H.P. (1996) Generalised confounding with Gröbner bases. *Biometrika* **83**, 653–666.

Pistone G. and Wynn H.P. (1999) Finitely generated cumulants. *Statist. Sinica* **9**, 1029–1052.

Riccomagno E. (1997) *Algebraic Geometry in Experimental Design and Related Fields*. PhD thesis, Department of Statistics, University of Warwick.

Robbiano L. (1985) Term orderings on the polynomial ring. In *EUROCAL '85, Vol. 2 (Linz, 1985)*, Springer, Berlin, 513–517.

Robbiano L. (1998) Gröbner bases and statistics. In *Gröbner Bases and Applications (Linz, 1998)*, Cambridge Univ. Press, Cambridge, 179–204.

Robbiano L. and Rogantin M.P. (1998) Full factorial designs and distracted fractions. In *Gröbner Bases and Applications (Linz, 1998)*, Cambridge Univ. Press, Cambridge, 473–482.

Sturmfels B. (1996) *Gröbner Bases and Convex Polytopes*. American Mathematical Society, Providence, RI.

Traverso C. (1996) Hilbert functions and the Buchberger algorithm. *J. Symbolic Comput.* **22**, 355–376.

Index

$\mathcal{L}(D, \mathbb{K})$, 96
CoCoA, 5, 46, 55

Abstract simplicial complex, 86
    Dimension, 86
Abstract Tube, 86

Boolean Algebra, 76

Coefficient Field, 6
Conditional Expectation, 108
Contrast, 92, 93

Design, 43
    2^d-full factorial, 75, 90–94
    3^{4-2}-fractional factorial, 48, 57
    Design Ideal, 26
    Echelon, 49
    Fraction, 75
    Full Factorial, 48, 56
    Generalized Echelon, 63
    Image, 48
    Maximal Fan, 64
    Minimal Fan, 63
    Product, 47, 57
    Restriction, 47
    Star Composite, 61
    Subset of a design, 66
    Union, 47, 67
Design Matrix, 1, 51, 97

Estimable Set, 51–59
Estimating Function, 139
Expectation, 101, 102, 104
Experimental Design, 1
Exponentials, 111

Generating Function: Cumulant, 114
Generating Function: Geometric, 114
Gröbner Basis, 27–30
    Buchberger Algorithm, 38
    Gauss Jordan Elimination, 39
    Minimal, 29
    of a design ideal, 44–47
    Reduced, 29
    Remainder, 29
    Total, 48
    total, 29
Grading, 21

Hilbert Basis Theorem, 24
    Dickson's Lemma, 24
Hilbert Function, 35, 65

Ideal
    homogeneous toric, 150
Identifiability, 53–56
Image Probability, 106
Image Support, 107
Inclusion-exclusion formula, 83
Indicator Function, *see* Interpolatory Polynomial, 77
Interpolation, 55
Interpolatory Polynomial, 50, 52, 84

Joint Probability Law, 106

Leading Coefficient, 22
Leading Monomial, 22
Leading Term, 22
Logic, 80

Maple, 5, 55

Mean Value, 91
Model, 15–17
    Confounding, 54–56
    Fourier Regression Model, 72
    Model Ideal, 26
    Parameter, 15
    Regression Model, 51, 71
Moment Aliasing, 101, 115
Moments, 3, 101
Monomial, see Term

Normal Form, see Remainder
Nullstellensatz Theorems
    Strong, 26
    Weak, 31

Order Ideal, 51, 54
Orthogonality
    in design theory, 92
    in exponential lattice models, 149

Parameter Confounding, 59
Parameters
    μ (moment), 121
    θ (polynomial), 121
    p (probability), 120
    mean, 138
    variance, 138
Polynomial, 16
    Orthogonal Polynomial, 71
    Polynomial Division, 2, 22–23, 36
    Polynomial Function, 16, 33
    Polynomial Ring, 15
    S-poly, 38
Polynomial Ideal, 17–19
    Design Ideal, 43–48
    Elimination, 18, 31, 45
    Fan, 60–66
        Leaf, 60
    generated by a set, 25
    Initial Ideal, 51, 60
    Monomial, 23
    Radical, 18, 26
Probability density, 103–105
Probability Event, 97

Quotient Space, 33–35
    Identifiability, 53–56

Random Variable, 97
Remainder, 2, 22, 23
    Gröbner Basis, 29
    Interpolation, 46
Ring
    of parametric functions, 126

Statistical Model
    Algebraic, 120
    Mixture, 122
Sufficient
    σ-algebra, 126
    ring, 126
    statistic, 124, 126
    statistics, 112
Support, 96
Support Ideal, 96
Support Matrix, see Design Matrix
Symmetric Difference, 92
System, 81
    Coherent, 81
    Failure Event, 81
    Failure Outcome, 81
    Outcome, 81
        Minimal Failure, 82
        Minimal Non-Failure, 82

Term, 16
    Log function, 16
Term-ordering, 19–22
    lex, 20
    tdeg, 20
    Block, 22, 72
    Elimination, 20
    Graded Ordering, 36
    Initial Ordering, 15, 19
    Restriction, 47
Tube theory, 83

Uniform Probability, 101

Variety, 16, 25
    generated by an ideal, 25